PERMACULTURE GARDENING FOR BEGINNERS

Jesse weber

Copyright © 2023 by Jesse weber

All rights reserved. No part of this book may be reproduced, distributed, or transmitted in any form or by any means, including photocopying, recording, or other electronic or mechanical methods, without the prior written permission of the publisher, except in the case of brief quotations embodied in critical reviews and certain other noncommercial uses permitted by copyright law.

Table of Content

Introduction	**9**
CHAPTER 1	**15**
What is Permaculture?	15
Definition and Principles	25
The History and Evolution of Permaculture	37
CHAPTER 2	**52**
Environmental Impact	64
Personal Benefits	77
CHAPTER 1	**86**
Getting Started	86
1. Understanding Your Space	100
Observing and Assessing Your Garden Area	113
Climate and Microclimates	125
2. Design Principles	137
Zoning and Sector Analysis	151
Designing for Sustainability	161
CHAPTER 2:	**173**
Soil Health	173
1. Building Healthy Soil	185
Composting Basics	198
Mulching and Cover Crops	211
2. Soil Testing and Amendments	224
Understanding Soil Types	240
Organic Soil Amendments	254
CHAPTER 3	**278**
Water Management	279

1. Harvesting and Conserving Water	295
Rainwater Harvesting	311
Greywater Systems	325
2. Irrigation Techniques	340
Drip Irrigation	356
Swales and Keyline Design	370
CHAPTER 4	**384**
Plant Selection	384
1. Choosing the Right Plants	397
Native and Adaptable Species	410
Companion Planting	422
2. Perennial vs. Annual Plants	433
Benefits of Perennials	445
Integrating Annuals into Your Garden	456
CHAPTER 5	**467**
Garden Layout and Design	467
1. Creating Plant Guilds	487
Examples of Plant Guilds	498
Benefits of Polycultures	509
2. Edible Landscapes	521
Integrating Food Forests	535
CHAPTER 6	**546**
Sustainable Practices	546
1. Natural Pest Control	557
Encouraging Beneficial Insects	569
Organic Pest Management	577
2. Crop Rotation and Diversity	588
Preventing Soil Depletion	600

Enhancing Biodiversity	610
Tools and Resources	620
1. Essential Tools for Permaculture Gardening	633
Must-Have Tools	646
Maintenance and Care	662
2. Further Reading and Learning	673
Recommended Books, Websites, and Courses	688
CHAPTER 8	**697**
Case Studies and Examples	697
1. Success Stories	709
Interviews with Experienced Permaculture Gardeners	721
Real-Life Examples of Permaculture Gardens	732

Introduction

Introduction to Permaculture Gardening for Beginners

What is Permaculture?

Permaculture is an innovative framework for creating sustainable ways of living. It is a practical method of developing ecologically harmonious, efficient, and productive systems that can be used by anyone, anywhere. Rooted in the principles of natural ecosystems, permaculture aims to integrate human habitats and agricultural systems with the natural environment.

The term "permaculture" was coined in the 1970s by Bill Mollison and David Holmgren, combining "permanent agriculture" and "permanent culture." This holistic approach emphasizes working with nature rather than against it, and seeks to create systems that are self-sustaining and resilient over time.

Core Principles of Permaculture

1. Observe and Interact: Take the time to engage with nature and understand its dynamics. By observing natural patterns and ecosystems, we can design more effective and sustainable human habitats.

2. Catch and Store Energy: Harness renewable resources like sunlight, wind, and rainwater to create energy stores that can be utilized when needed.

3. Obtain a Yield: Ensure that your systems are producing outputs that are beneficial, such as food, water, or energy.

4. Apply Self-Regulation and Accept Feedback: Incorporate feedback loops to help systems self-regulate and become more resilient over time.

5. Use and Value Renewable Resources and Services: Reduce dependence on non-renewable resources by utilizing natural systems and renewable materials.

6. Produce No Waste: Design systems to use outputs from one process as inputs for another, minimizing waste and maximizing efficiency.

7. Design From Patterns to Details: Observe natural patterns and use them as the foundation for your design, filling in the details as you go.

8. Integrate Rather Than Segregate: Create connections between elements in your system to promote synergy and mutual support.

9. Use Small and Slow Solutions: Start small and scale up gradually, allowing systems to evolve naturally and sustainably.

10. Use and Value Diversity: Embrace a variety of plants, animals, and practices to enhance system resilience and productivity.

11. Use Edges and Value the Marginal: The edges of systems are often the most productive and diverse areas; value and utilize these spaces.

12. Creatively Use and Respond to Change: Be flexible and adaptable, allowing your system to evolve in response to changing conditions.

Why Permaculture Gardening?

Permaculture gardening offers a multitude of benefits, both for individuals and for the environment. Here are some compelling reasons to adopt this approach:

1. Environmental Impact:

- **Soil Health:** Permaculture techniques, such as composting and mulching, enhance soil fertility and structure, leading to healthier plants and reduced erosion.
- **Biodiversity:** By promoting polycultures and diverse plant guilds, permaculture gardens support a wide range of species, creating habitats for beneficial insects, birds, and other wildlife.
- **Water Conservation:** Techniques like rainwater harvesting, greywater systems, and swales help to capture and store water, reducing the need for irrigation and mitigating the impact of drought.
- **Climate Change Mitigation:** Permaculture gardens sequester carbon in

the soil and reduce greenhouse gas emissions by minimizing the use of synthetic fertilizers and pesticides.

2. Personal Benefits:
   - Food Security: Growing your own food ensures a steady supply of fresh, nutritious produce, reducing dependence on external food sources.
   - **Health and Well-Being:** Gardening is a physical activity that promotes mental and physical health, providing a sense of accomplishment and connection to nature.
   - **Cost Savings:** By producing your own food, you can save money on groceries and reduce household expenses.
   - **Community Building**: Permaculture gardens often foster community connections, as neighbors share resources, knowledge, and surplus produce.

The Journey Ahead
Embarking on a journey into permaculture gardening is an exciting and rewarding

endeavor. Whether you have a small balcony or a large backyard, the principles of permaculture can be applied to create a thriving, sustainable garden that benefits both you and the environment.

In this book, we will guide you through the process of understanding your space, designing your garden, building healthy soil, managing water, selecting plants, and integrating sustainable practices. Along the way, we will provide practical tips, real-life examples, and inspiring case studies to help you succeed in your permaculture gardening journey.

Remember, the essence of permaculture lies in observing, learning, and adapting. As you work with nature, you will discover the joy and satisfaction of creating a garden that is not only beautiful and productive but also resilient and harmonious with the environment. Welcome to the world of permaculture gardening—let's get started!

CHAPTER 1

What is Permaculture?

Origins and Evolution

Permaculture, a term coined in the 1970s by Bill Mollison and David Holmgren, stands for "permanent agriculture" and "permanent culture." It emerged from the necessity to create sustainable agricultural systems that harmonize with natural ecosystems. This innovative approach integrates various disciplines, including agriculture, architecture, ecology, and landscape design, to develop self-sufficient, resilient, and regenerative systems.

Bill Mollison, an Australian ecologist, and David Holmgren, an environmental designer, developed permaculture principles in response to the destructive agricultural practices that dominated the 20th century. Their vision was to create agricultural systems that mimic the diversity, stability, and resilience of natural ecosystems. Over the years, permaculture has evolved into a holistic philosophy applicable not just to farming, but to all aspects of human habitation.

Core Principles of Permaculture

Permaculture is guided by a set of ethics and principles that inform its design practices. These principles are divided into three core ethics and twelve design principles, all of which work together to create sustainable and regenerative systems.

Core Ethics:
1. Earth Care: Protecting and regenerating natural systems, ensuring the health of the planet and its ecosystems. This involves practices that restore soil fertility, conserve water, and enhance biodiversity.

2. People Care: Ensuring that human needs are met in ways that promote well-being and social justice. This includes building communities, enhancing local economies, and creating environments that support human health and happiness.

3. Fair Share: Distributing surplus in a way that supports Earth Care and People Care. This means sharing resources, knowledge, and time to create equitable systems that do not over-exploit the environment or people.

Design Principles:

1. Observe and Interact: Careful observation of natural systems and human interactions provides valuable insights for designing sustainable solutions. Engage with nature to understand its patterns and processes.

2. Catch and Store Energy: Harness renewable resources like sunlight, water, and wind, storing them for future use. This can include solar panels, rainwater harvesting systems, and energy-efficient building designs.

3. Obtain a Yield: Ensure that your systems produce outputs that are beneficial, such as food, energy, or materials. Every element should contribute to the productivity of the system.

4. Apply Self-Regulation and Accept Feedback: Incorporate feedback loops to help systems self-regulate and become more resilient. Learn from mistakes and adapt your practices accordingly.

5. Use and Value Renewable Resources and Services: Reduce dependence on non-renewable resources by utilizing natural systems and renewable materials. This can include organic farming methods, renewable energy sources, and sustainable materials.

6. Produce No Waste: Design systems to use outputs from one process as inputs for another, minimizing waste and maximizing efficiency. Composting, recycling, and reusing materials are key strategies.

7. Design From Patterns to Details: Observe natural patterns and use them as the foundation for your design, filling in the details as you go. Understanding the larger patterns in nature can inform the layout and function of your systems.

8. Integrate Rather Than Segregate: Create connections between elements in your system to promote synergy and mutual support. Plants, animals, and structures should work together to enhance the overall system.

9. Use Small and Slow Solutions: Start small and scale up gradually, allowing systems to evolve naturally and sustainably. Small, manageable changes are more likely to succeed and be sustainable.

10. Use and Value Diversity: Embrace a variety of plants, animals, and practices to enhance system resilience and productivity. Biodiversity creates more resilient and productive ecosystems.

11. Use Edges and Value the Marginal: The edges of systems are often the most productive and diverse areas; value and utilize these spaces. This can include the edges of ponds, forests, and fields.

12. Creatively Use and Respond to Change: Be flexible and adaptable, allowing your system to evolve in response to changing conditions. Change is inevitable, and successful systems are those that can adapt and thrive.

Practical Applications of Permaculture

Permaculture principles can be applied to a wide range of settings and activities, from small urban gardens to large-scale agricultural projects. Here are some key areas where permaculture can make a significant impact:

1. Gardening and Agriculture:
   - **Polyculture and Companion Planting:** Growing multiple crops together in a way that they benefit each other, enhancing biodiversity and reducing pest problems.

- **Food Forests:** Designing multi-layered gardens that mimic natural forests, providing diverse yields of fruits, nuts, vegetables, and herbs.
- **Soil Building:** Techniques such as composting, mulching, and cover cropping to improve soil fertility and structure.
- **Water Management:** Implementing swales, ponds, and rainwater harvesting systems to conserve and manage water resources effectively.

2. Architecture and Urban Design:
- Energy-Efficient Buildings: Designing homes and buildings that utilize passive solar heating, natural cooling, and energy-efficient construction techniques.
- **Green Infrastructure:** Incorporating green roofs, vertical gardens, and urban agriculture to enhance urban environments and provide food and habitat.
- **Community Planning:** Creating walkable, mixed-use neighborhoods that

foster social interaction and reduce reliance on automobiles.

3. Resource Management:
   - **Renewable Energy Systems:** Utilizing solar, wind, and other renewable energy sources to reduce dependency on fossil fuels.
   - **Waste Reduction:** Designing systems that minimize waste through recycling, composting, and the repurposing of materials.
   - **Sustainable Transportation:** Promoting walking, cycling, public transportation, and other low-impact transportation options.

4. Social Systems:
   - **Community Building:** Fostering strong, resilient communities through cooperative housing, shared resources, and collaborative decision-making.
   - **Education and Outreach:** Teaching permaculture principles and practices

through workshops, courses, and community projects.

- **Economic Systems:** Supporting local economies through cooperatives, local currencies, and sustainable business practices.

The Global Impact of Permaculture

Permaculture has the potential to address some of the most pressing challenges facing humanity, including climate change, food security, and resource depletion. By promoting sustainable and regenerative practices, permaculture can help create a world where humans live in harmony with nature, and where our needs are met without compromising the ability of future generations to meet theirs.

Permaculture projects and initiatives are thriving around the world, from urban gardens in New York City to reforestation efforts in sub-Saharan Africa. These projects demonstrate the power of permaculture to transform communities, restore ecosystems, and create resilient, sustainable systems.

Conclusion

Permaculture is more than just a method of gardening; it is a holistic approach to living sustainably and harmoniously with the natural world. By understanding and applying its core principles, we can create systems that are not only productive and efficient but also regenerative and resilient. Whether you are a beginner gardener or an experienced farmer, permaculture offers a framework for creating a better, more sustainable future.

Definition and Principles

Permaculture is a design philosophy and system that creates sustainable, self-sufficient, and regenerative ecosystems. It is rooted in the observation of natural systems and seeks to mimic their diversity, resilience, and efficiency. By integrating human habitats and agricultural systems with the natural environment, permaculture aims to achieve a harmonious balance that supports both human and ecological well-being.

The term "permaculture" was coined by Bill Mollison and David Holmgren in the 1970s, combining the words "permanent" and "agriculture" to signify the creation of agricultural systems that are sustainable and enduring. Over time, the concept has expanded to include "permanent culture," reflecting the broader application of permaculture principles to all aspects of human life, including housing, community planning, and resource management.

Core Ethics of Permaculture

Permaculture is guided by three core ethics that form the foundation of its philosophy:

1. Earth Care: This ethic emphasizes the importance of preserving and regenerating natural systems. It involves practices that enhance soil health, conserve water, protect biodiversity, and reduce environmental degradation. Earth Care recognizes that a healthy planet is essential for sustaining human life and other forms of life.

2. People Care: People Care focuses on meeting the basic needs of individuals and communities in ways that promote well-being and social justice. This includes creating environments that support physical and mental health, fostering community connections, and ensuring access to resources such as clean water, healthy food, and safe shelter.

3. Fair Share: Also known as "Return of Surplus," this ethic encourages the equitable distribution of resources and the sharing of surplus. It involves using only what we need and redistributing excess to support Earth Care and People Care. Fair Share promotes the idea of limits to consumption and advocates for sharing resources, knowledge, and time to create more just and sustainable societies.

Twelve Principles of Permaculture

Permaculture design is guided by twelve principles that provide a practical framework for applying its ethics. These principles, developed by David Holmgren, are based on the observation of natural systems and aim to create efficient, resilient, and sustainable designs.

1. Observe and Interact:
   - Take the time to engage with nature and observe its patterns and processes. By understanding the environment and its dynamics, we can design more effective and sustainable solutions.
   - **Example**: Before planting a garden, observe the site to understand sunlight patterns, soil conditions, and existing vegetation.

2. Catch and Store Energy:
   - Harness renewable resources like sunlight, wind, and rainwater, and store them for future use. This helps create a stable and reliable source of energy.
   - **Example:** Installing solar panels to generate electricity and using rain barrels to collect rainwater for irrigation.

3. Obtain a Yield:
   - Ensure that your systems produce outputs that are beneficial, such as food,

energy, or materials. Every element in the system should contribute to its productivity.

   - **Example:** Growing vegetables and fruits in a garden to provide food for the household.

4. Apply Self-Regulation and Accept Feedback:

   - Incorporate feedback loops to help systems self-regulate and become more resilient. Learn from mistakes and adapt your practices accordingly.

   - **Example**: Monitoring soil health and adjusting composting practices based on observed deficiencies.

5. Use and Value Renewable Resources and Services:

   - Reduce dependence on non-renewable resources by utilizing natural systems and renewable materials. This can include organic farming methods, renewable energy sources, and sustainable materials.

- **Example:** Using compost instead of synthetic fertilizers to enrich soil.

6. Produce No Waste:
- Design systems to use outputs from one process as inputs for another, minimizing waste and maximizing efficiency.
- **Example:** Composting kitchen scraps and garden waste to create nutrient-rich soil amendments.

7. Design From Patterns to Details:
- Observe natural patterns and use them as the foundation for your design, filling in the details as you go. Understanding the larger patterns in nature can inform the layout and function of your systems.
- Example: Designing a garden layout based on the natural flow of water across the site.

8. Integrate Rather Than Segregate:
- Create connections between elements in your system to promote synergy and mutual

support. Plants, animals, and structures should work together to enhance the overall system.

   - **Example:** Planting companion plants that benefit each other, such as beans and corn.

9. Use Small and Slow Solutions:
   - Start small and scale up gradually, allowing systems to evolve naturally and sustainably. Small, manageable changes are more likely to succeed and be sustainable.
   - **Example**: Starting with a small garden bed and expanding as you gain experience and confidence.

10. Use and Value Diversity:
   - Embrace a variety of plants, animals, and practices to enhance system resilience and productivity. Biodiversity creates more resilient and productive ecosystems.

- **Example**: Planting a diverse range of crops to reduce the risk of pests and diseases.

11. Use Edges and Value the Marginal:
- The edges of systems are often the most productive and diverse areas; value and utilize these spaces.

- **Example**: Planting along the edges of a pond or creating hedgerows that provide habitat for beneficial insects and wildlife.

12. Creatively Use and Respond to Change:
- Be flexible and adaptable, allowing your system to evolve in response to changing conditions. Change is inevitable, and successful systems are those that can adapt and thrive.

- **Example:** Adjusting planting schedules and crop selections in response to climate variations or pest pressures.

Applying Permaculture Principles

Permaculture principles can be applied to various aspects of human life, from gardening and agriculture to architecture, urban planning, and resource management. Here are some practical applications of these principles:

1. Gardening and Agriculture:
  - **Polyculture and Companion Planting:** Growing multiple crops together in a way that they benefit each other, enhancing biodiversity and reducing pest problems.
  - **Food Forests:** Designing multi-layered gardens that mimic natural forests, providing diverse yields of fruits, nuts, vegetables, and herbs.
   - **Soil Building:** Techniques such as composting, mulching, and cover cropping to improve soil fertility and structure.
    - **Water Management:** Implementing swales, ponds, and rainwater harvesting

systems to conserve and manage water resources effectively.

2. Architecture and Urban Design:
- **Energy-Efficient Buildings**: Designing homes and buildings that utilize passive solar heating, natural cooling, and energy-efficient construction techniques.
- **Green Infrastructure**: Incorporating green roofs, vertical gardens, and urban agriculture to enhance urban environments and provide food and habitat.
- **Community Planning:** Creating walkable, mixed-use neighborhoods that foster social interaction and reduce reliance on automobiles.

3. Resource Management:
- **Renewable Energy Systems:** Utilizing solar, wind, and other renewable energy sources to reduce dependency on fossil fuels.
- **Waste Reduction:** Designing systems that minimize waste through recycling,

composting, and the repurposing of materials.

   - **Sustainable Transportation:** Promoting walking, cycling, public transportation, and other low-impact transportation options.

4. Social Systems:
   - **Community Building:** Fostering strong, resilient communities through cooperative housing, shared resources, and collaborative decision-making.
   - **Education and Outreach:** Teaching permaculture principles and practices through workshops, courses, and community projects.
   - **Economic Systems:** Supporting local economies through cooperatives, local currencies, and sustainable business practices.

Conclusion

Permaculture is a holistic design philosophy that seeks to create sustainable, self-sufficient, and regenerative ecosystems by mimicking the diversity, resilience, and efficiency of natural systems. Guided by its core ethics of Earth Care, People Care, and Fair Share, and its twelve design principles, permaculture provides a practical framework for designing systems that are not only productive and efficient but also resilient and adaptable.

Whether applied to gardening, architecture, urban planning, or resource management, permaculture offers solutions that enhance ecological health, promote social equity, and support sustainable living. By embracing permaculture principles, individuals and communities can create a more sustainable and harmonious world, where human needs are met in ways that respect and regenerate the natural environment.

The History and Evolution of Permaculture

The History and Evolution of Permaculture

Early Influences and Roots

The roots of permaculture can be traced back to traditional agricultural practices and indigenous knowledge systems that have been used for centuries. These practices were characterized by a deep understanding of local ecosystems and a sustainable approach to land use that sought to maintain ecological balance. Traditional techniques such as polyculture, crop rotation, and the use of organic fertilizers were common in many indigenous cultures around the world.

The formal development of permaculture as a distinct concept began in the mid-20th century, influenced by the growing environmental awareness and the countercultural movements of the 1960s and 1970s. This period saw a rise in ecological thinking, as well as a critique of industrial agriculture and its negative impacts on the environment.

The Birth of Permaculture: Bill Mollison and David Holmgren

Permaculture as we know it today was co-founded by Bill Mollison and David Holmgren in the 1970s. Bill Mollison, an Australian ecologist, and David Holmgren, an environmental designer and student of Mollison, developed the initial principles of permaculture as a response to the environmental crises they observed.

1. Bill Mollison:
   - Born in Tasmania, Australia, in 1928, Mollison's early experiences as a fisherman and wildlife researcher shaped his understanding of natural systems. His observations of the degradation caused by industrial agriculture motivated him to seek sustainable alternatives.
   - Mollison's teaching career at the University of Tasmania provided a platform for his ecological ideas. In 1974, he began

collaborating with David Holmgren to develop the concept of permaculture.

2. David Holmgren:

- Holmgren was a student of environmental design and a protégé of Mollison. Together, they published "Permaculture One" in 1978, a seminal work that outlined the principles of permaculture and introduced the concept to a wider audience.

Development and Spread of Permaculture

Following the publication of "Permaculture One," Mollison and Holmgren continued to refine and promote the principles of permaculture. Mollison, in particular, became a prominent advocate, teaching courses and writing extensively on the subject.

1. Permaculture Design Course (PDC):
   - In the early 1980s, Mollison developed the Permaculture Design Course (PDC), a comprehensive curriculum that provided participants with the knowledge and skills to design sustainable systems. The PDC became a cornerstone of the permaculture movement, and thousands of people worldwide have since completed the course.

2. International Expansion:
   - Mollison's extensive travels and teaching tours helped spread permaculture beyond Australia. He taught courses in North America, Europe, Africa, and Asia, adapting permaculture principles to diverse climates and cultures.
   - Publications such as "Permaculture: A Designer's Manual" (1988) and "Introduction to Permaculture" (1991) further disseminated permaculture knowledge and practices.

3. Local and Regional Adaptations:

- As permaculture spread globally, practitioners began adapting the principles to local conditions. This flexibility allowed permaculture to thrive in a variety of environments, from temperate climates to tropical and arid regions.

Evolution of Permaculture Concepts and Practices

Over the decades, permaculture has evolved to encompass a broad range of concepts and practices, reflecting its adaptability and holistic approach. Key developments include:

1. Broadening the Scope:

- Initially focused on sustainable agriculture and land use, permaculture has expanded to include urban design, architecture, community planning, and social systems. This broader scope emphasizes the interconnectedness of ecological, economic, and social sustainability.

2. Integration of New Technologies:
- While rooted in traditional knowledge, permaculture has also embraced modern technologies and scientific advancements. Innovations such as renewable energy systems, aquaponics, and regenerative agriculture techniques have been integrated into permaculture practices.

3. Focus on Community and Social Systems:
- The social dimension of permaculture has gained increasing attention. Projects now often include elements of community building, cooperative economics, and education, recognizing that sustainable living requires strong, resilient communities.

4. Research and Education:
- Permaculture research institutions and educational centers have been established

worldwide, contributing to the ongoing development and dissemination of permaculture knowledge. These institutions conduct research, offer training programs, and support the implementation of permaculture projects.

Significant Figures and Influences

Several individuals and movements have significantly influenced the development and spread of permaculture:

1. Masanobu Fukuoka:
   - A Japanese farmer and philosopher, Fukuoka's "do-nothing" approach to farming, as outlined in his book "The One-Straw Revolution" (1975), inspired permaculture practitioners. His emphasis on minimal intervention and working with nature resonated with permaculture principles.

2. Sepp Holzer:

- An Austrian farmer known for his innovative permaculture practices, Holzer developed a highly productive farm in the challenging alpine environment. His work has demonstrated the potential of permaculture to create resilient and diverse agricultural systems in difficult conditions.

3. Geoff Lawton:
- A prominent permaculture teacher and designer, Lawton has played a key role in spreading permaculture through his teaching and online presence. His instructional videos and courses have reached a global audience, further popularizing permaculture principles.

Contemporary Permaculture Movements and Projects

Today, permaculture is a global movement with a diverse array of projects and initiatives that demonstrate its principles in action. Some notable examples include:

1. Urban Permaculture:
   - Cities around the world are adopting permaculture principles to create green infrastructure, urban gardens, and community food systems. Examples include urban farms in Detroit, community gardens in London, and green rooftops in New York City.

2. Regenerative Agriculture:
   - Regenerative agriculture practices, which aim to restore soil health, enhance biodiversity, and sequester carbon, are closely aligned with permaculture. Farms and ranches worldwide are adopting regenerative practices to create sustainable food systems.

3. Ecovillages and Intentional Communities:

- Ecovillages and intentional communities, such as Auroville in India and Findhorn in Scotland, are applying permaculture principles to create sustainable living environments. These communities serve as living laboratories for permaculture practices and social innovation.

4. Disaster Resilience and Relief:
- Permaculture is being used to create resilient systems in areas affected by natural disasters and humanitarian crises. Projects in places like Haiti, Nepal, and Puerto Rico are helping communities rebuild using sustainable and regenerative practices.

The Future of Permaculture

As the global challenges of climate change, resource depletion, and social inequality become more pressing, the principles and practices of permaculture are more relevant than ever. The future of permaculture lies in its continued evolution and adaptation to meet these challenges. Key areas of focus include:

1. Climate Change Adaptation and Mitigation:
- Permaculture can play a vital role in helping communities adapt to and mitigate the impacts of climate change. Techniques such as agroforestry, water management, and carbon sequestration are essential tools for building climate resilience.

2. Food Security and Sovereignty:
- By promoting local food production and sustainable agriculture, permaculture can enhance food security and sovereignty. Community-supported agriculture (CSA), local food networks, and permaculture farms contribute to resilient food systems.

3. Education and Capacity Building:

- Expanding permaculture education and training programs is crucial for spreading knowledge and empowering individuals and communities to implement sustainable practices. Online courses, workshops, and community-based training programs are essential for reaching a broader audience.

4. Policy and Advocacy:
- Engaging with policymakers and advocating for supportive policies is necessary to scale up permaculture practices. This includes promoting land use policies, funding for sustainable agriculture, and incentives for renewable energy adoption.

5. Innovation and Research:
- Continued innovation and research are vital for advancing permaculture. Collaborative research projects, experimentation with new techniques, and integration of traditional knowledge with

modern science will drive the evolution of permaculture practices.

Conclusion

The history and evolution of permaculture reflect a dynamic and adaptive approach to creating sustainable, self-sufficient, and regenerative systems. From its roots in traditional knowledge and ecological observation to its global spread and contemporary applications, permaculture has become a powerful tool for addressing some of the most pressing challenges of our time.

By embracing the core ethics of Earth Care, People Care, and Fair Share, and applying the twelve design principles, individuals and communities can create resilient systems that enhance ecological health, promote social equity, and support sustainable living. As we look to the future, the continued evolution and adaptation of permaculture will be essential for building a more sustainable and harmonious world.

CHAPTER 2

Introduction

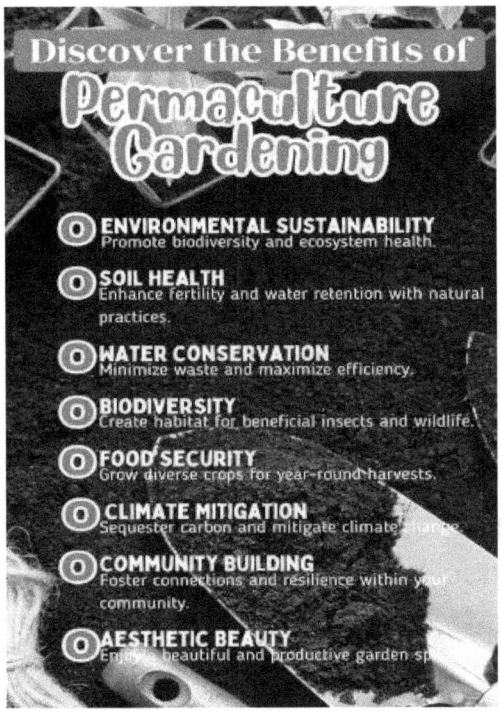

Permaculture gardening offers a holistic approach to sustainable living that extends far beyond traditional gardening practices. By mimicking natural is ecosystems, permaculture gardens create self-sustaining, resilient, and productive environments that require less external input and yield more diverse and abundant outputs. This approach addresses many of the pressing challenges facing modern agriculture and urban living, from environmental degradation and climate change to food security and community resilience.

Environmental Benefits

1. Soil Health and Regeneration:
   - Traditional agriculture often depletes soil nutrients, leading to erosion, loss of fertility, and a decline in soil structure. Permaculture gardening emphasizes practices such as composting, mulching, and crop rotation, which build soil health over time.
   - Techniques like no-till gardening and cover cropping prevent soil erosion and improve soil structure, fostering a rich, living soil that supports plant growth and sequesters carbon.

2. Water Conservation and Management:
   - Permaculture gardens are designed to maximize water efficiency. Practices like rainwater harvesting, swales, and keyline design capture and store rainwater, reducing the need for irrigation and preventing runoff and erosion.
   - Mulching and polyculture planting reduce water evaporation from the soil, maintaining moisture levels and promoting healthy plant growth.

3. Biodiversity and Ecosystem Health:
   - By incorporating a diverse range of plants, animals, and microorganisms, permaculture gardens create robust ecosystems that are more resilient to pests, diseases, and extreme weather events.
   - Planting a variety of species attracts beneficial insects, birds, and pollinators, enhancing biodiversity and ecological balance. This reduces the need for chemical

pesticides and fertilizers, promoting a healthier environment.

4. Carbon Sequestration and Climate Resilience:
   - Permaculture practices such as agroforestry, cover cropping, and organic soil management sequester carbon in the soil and vegetation, helping mitigate climate change.
   - Diverse, multi-layered planting systems provide shade, reduce heat islands in urban areas, and create microclimates that buffer against extreme weather conditions.

Economic Benefits

1. Cost Savings and Resource Efficiency:
   - Permaculture gardens require fewer external inputs like synthetic fertilizers, pesticides, and water, resulting in significant cost savings over time.
   - Utilizing local and renewable resources, such as compost and rainwater, reduces

dependence on expensive, non-renewable inputs.

2. Increased Yields and Food Security:
- Polyculture and companion planting strategies in permaculture gardens lead to higher yields per unit area compared to monoculture systems. This diversity of crops ensures a steady supply of fresh produce throughout the year.
- Growing a variety of food crops in a small space increases food security, reducing reliance on external food sources and supply chains that can be disrupted by economic or environmental crises.

3. Local Economy and Job Creation:
- Permaculture gardening supports local economies by encouraging the production and sale of locally grown food. This fosters community self-reliance and reduces the environmental impact of transporting food over long distances.

- Permaculture projects can create jobs in areas such as garden design, education, and local food production, contributing to economic resilience and community well-being.

Social and Community Benefits

1. Community Building and Collaboration:
   - Permaculture gardening often involves community participation and collaboration, fostering a sense of belonging and shared purpose. Community gardens, urban farms, and cooperative projects bring people together, strengthening social ties and networks.
   - Educational programs and workshops on permaculture principles and practices empower individuals with the knowledge and skills to contribute to community sustainability.

2. Health and Well-Being:

- Growing and consuming fresh, organic produce from a permaculture garden improves nutritional intake and promotes better health outcomes. Gardening activities also provide physical exercise and reduce stress, contributing to overall well-being.

- Access to green spaces and nature has been shown to improve mental health, reduce anxiety and depression, and enhance cognitive function.

3. Education and Empowerment:

- Permaculture gardening serves as a practical tool for environmental education, teaching people about sustainable living, ecological principles, and resource management. Schools, community centers, and educational institutions can use permaculture gardens as living laboratories for experiential learning.

- Empowering individuals with the skills to grow their own food and manage natural resources sustainably fosters self-reliance and resilience.

Practical and Aesthetic Benefits

1. Low Maintenance and Self-Sustaining Systems:
   - Once established, permaculture gardens require less maintenance than conventional gardens. The emphasis on perennials, self-seeding annuals, and natural pest control reduces the need for constant intervention.
   - By designing systems that mimic natural ecosystems, permaculture gardens create self-sustaining environments where plants, animals, and microorganisms work together in harmony.

2. Aesthetic and Recreational Value:

- Permaculture gardens are not only functional but also aesthetically pleasing. The diverse plantings, natural structures, and wildlife habitats create beautiful and serene spaces that enhance the quality of life.
- These gardens provide recreational opportunities for individuals and families, offering a space for relaxation, play, and connection with nature.

Ethical and Philosophical Benefits

1. Ethical Stewardship and Responsibility:
- Permaculture gardening embodies the ethics of Earth Care, People Care, and Fair Share. It encourages responsible stewardship of the land, respect for all living beings, and a commitment to sustainability.
- By practicing permaculture, individuals and communities contribute to the regeneration of the planet and the well-being of future generations.

2. Holistic and Systems Thinking:
- Permaculture promotes holistic and systems thinking, encouraging a deeper understanding of the interconnectedness of natural and human systems. This perspective fosters innovative and integrative solutions to complex problems.
- Adopting a permaculture mindset can lead to more conscious and intentional living, where decisions are made with consideration of their broader ecological and social impacts.

Global Impact and Sustainability

1. Addressing Global Challenges:
- Permaculture gardening offers practical solutions to global challenges such as climate change, food insecurity, and environmental degradation. By creating resilient and sustainable systems, permaculture contributes to global efforts to achieve sustainability goals.

- Permaculture principles can be adapted to diverse environments and cultures, making it a versatile and globally relevant approach to sustainability.

2. Inspiring Positive Change:

- The success of permaculture projects around the world serves as a powerful example of what is possible when people work in harmony with nature. These projects inspire others to adopt sustainable practices and contribute to the growing movement for ecological and social regeneration.

- Permaculture's emphasis on local action and community empowerment encourages grassroots initiatives and collective action, driving positive change from the ground up.

Conclusion

Permaculture gardening offers a comprehensive and sustainable approach to creating productive, resilient, and beautiful environments. By embracing permaculture principles, individuals and communities can achieve a wide range of benefits, from environmental conservation and economic savings to social well-being and ethical stewardship.

In a world facing numerous ecological and social challenges, permaculture gardening provides a hopeful and practical path forward. It empowers people to take control of their food production, reduce their environmental footprint, and build stronger, healthier communities. As more individuals and communities adopt permaculture practices, the cumulative impact can drive significant progress towards a more sustainable and harmonious world.

Environmental Impact

Environmental Impact of Permaculture Gardening

Introduction

Permaculture gardening, with its emphasis on sustainability, ecological balance, and regenerative practices, has a profound and multifaceted impact on the environment. By mimicking natural ecosystems and prioritizing renewable resources, permaculture gardens contribute to soil health, water conservation, biodiversity, and climate resilience. This approach offers a sustainable alternative to conventional agricultural practices, which often lead to environmental degradation. In this comprehensive exploration, we will delve into the various environmental benefits of permaculture gardening.

Soil Health and Regeneration

1. Soil Structure and Fertility:
 - Permaculture gardening practices such as no-till gardening, composting, and mulching enhance soil structure and fertility. These methods reduce soil compaction, increase aeration, and promote the formation of humus, a key component of fertile soil.
 - By adding organic matter to the soil, permaculture gardens improve its ability to retain water and nutrients, supporting healthy plant growth and reducing the need for chemical fertilizers.

2. Erosion Prevention:
 - Techniques like contour planting, swales, and terracing are used in permaculture gardens to manage water flow and prevent soil erosion. These methods slow down water runoff, allowing it to infiltrate the soil rather than washing away topsoil.

- Cover crops and ground covers protect the soil surface from erosion by wind and rain, maintaining soil integrity and preventing nutrient loss.

3. Soil Microbiology:
 - Healthy soil is teeming with beneficial microorganisms such as bacteria, fungi, and nematodes. These microorganisms play a crucial role in nutrient cycling, organic matter decomposition, and disease suppression.
 - Permaculture practices that avoid synthetic chemicals and promote organic inputs support a thriving soil microbiome, enhancing soil health and plant resilience.

Water Conservation and Management

1. Efficient Water Use:
 - Permaculture gardens are designed to maximize water efficiency through techniques like rainwater harvesting, drip irrigation, and greywater recycling. These

methods reduce the need for external water sources and ensure that water is used efficiently and effectively.

- Mulching helps to retain soil moisture, reducing evaporation and the frequency of irrigation needed to maintain healthy plant growth.

2. Rainwater Harvesting:

- Capturing rainwater from rooftops and other surfaces provides a sustainable source of water for irrigation. Rain barrels, cisterns, and ponds can store rainwater, making it available during dry periods.

- Rainwater harvesting reduces the strain on municipal water supplies and decreases runoff that can lead to flooding and water pollution.

3. Swales and Keyline Design:

- Swales, or shallow trenches dug along the contour of the land, capture and direct rainwater into the soil, promoting groundwater recharge and reducing runoff. This technique helps to maintain soil moisture levels and supports plant growth.

- Keyline design is a land management practice that optimizes water distribution across a landscape. By strategically placing swales, ponds, and other water management features, keyline design enhances water infiltration and minimizes erosion.

Biodiversity and Ecosystem Health

1. Diverse Plantings:
   - Permaculture gardens incorporate a wide variety of plants, including annuals, perennials, shrubs, and trees. This diversity mimics natural ecosystems and creates a more resilient and productive garden.
   - Plant diversity supports a range of beneficial insects, birds, and other wildlife, promoting ecological balance and reducing the need for chemical pest control.

2. Habitat Creation:
   - By including elements such as ponds, hedgerows, and native plants, permaculture gardens provide habitat for wildlife. These features attract pollinators, birds, amphibians, and other beneficial creatures that contribute to a healthy ecosystem.
   - Habitat creation also supports biodiversity by offering food, shelter, and breeding sites for various species.

3. Integrated Pest Management:
   - Permaculture gardens use natural pest control methods that minimize or eliminate the need for synthetic pesticides. Techniques such as companion planting, attracting beneficial insects, and creating habitat for predators help manage pest populations.
   - For example, planting marigolds alongside vegetables can repel nematodes, while providing habitat for ladybugs helps control aphid populations.

Climate Resilience and Carbon Sequestration

1. Carbon Sequestration:
- Permaculture practices such as agroforestry, cover cropping, and organic soil management sequester carbon in both plants and soil. Trees and perennial plants absorb carbon dioxide from the atmosphere and store it as biomass, while healthy soils can capture and retain significant amounts of carbon.
- By enhancing soil organic matter through composting and mulching, permaculture gardens increase the soil's capacity to sequester carbon, mitigating the impacts of climate change.

2. Climate Resilience:
- Diverse, multi-layered planting systems in permaculture gardens create microclimates that buffer against extreme weather conditions. Shade provided by

trees and shrubs reduces temperature fluctuations and protects soil and plants from heat stress.

- The resilience of permaculture gardens to drought, floods, and other climate-related events is enhanced by practices that improve soil health, water management, and biodiversity.

Reduction of Pollution and Chemical Use

1. Elimination of Synthetic Chemicals:
- Permaculture gardening avoids the use of synthetic fertilizers, pesticides, and herbicides, which can harm the environment and human health. Instead, organic inputs such as compost, mulch, and natural pest repellents are used to maintain soil fertility and manage pests.
- Reducing chemical use prevents soil and water pollution, protects beneficial organisms, and promotes a healthier ecosystem.

2. Waste Reduction and Recycling:

- Permaculture emphasizes the use of local and renewable resources, minimizing waste and promoting recycling. Organic waste from the garden and kitchen can be composted and returned to the soil as valuable nutrients.
- Practices such as vermiculture (worm composting) and bokashi (fermented composting) provide efficient ways to recycle organic waste and enhance soil fertility.

Contribution to Sustainable Food Systems

1. Local Food Production:

- By growing food locally, permaculture gardens reduce the environmental impact of transporting food over long distances. This decreases greenhouse gas emissions associated with food transportation and supports local food systems.

- Local food production also enhances food security, providing fresh, nutritious produce close to where it is consumed.

2. Seasonal and Diverse Harvests:
- Permaculture gardens produce a variety of crops throughout the year, offering a steady supply of fresh food. This diversity reduces dependency on monoculture crops and enhances dietary variety and nutrition.
- Seasonal harvesting aligns with natural cycles, reducing the need for energy-intensive food preservation and storage.

Promoting Regenerative Agriculture

1. Regenerative Practices:
- Permaculture gardening aligns with regenerative agriculture principles, which focus on restoring and enhancing ecosystem health. Practices such as agroforestry, polyculture, and holistic

grazing regenerate soil, improve water cycles, and increase biodiversity.

- Regenerative agriculture goes beyond sustainability, aiming to heal and rebuild natural systems rather than merely maintaining them.

2. Long-Term Sustainability:

- By creating self-sustaining systems that require minimal external inputs, permaculture gardens ensure long-term sustainability. These systems are designed to be resilient, adaptable, and capable of regenerating themselves over time.

- The focus on local resources, ecological balance, and renewable inputs creates gardens that can thrive indefinitely without depleting natural resources.

Global and Local Environmental Impact

1. Scalability and Adaptability:

- Permaculture principles can be applied at various scales, from small urban gardens to large agricultural landscapes. This scalability makes permaculture accessible to individuals, communities, and farmers worldwide.

- The adaptability of permaculture practices to different climates and cultures enhances their global relevance and impact.

2. Community and Ecosystem Resilience:

- By promoting local food production, biodiversity, and sustainable resource management, permaculture gardens contribute to the resilience of both communities and ecosystems. This resilience is crucial in the face of environmental challenges such as climate change, resource depletion, and biodiversity loss.

- Permaculture projects often serve as models of sustainable living, inspiring and educating others to adopt similar practices and contribute to broader environmental goals.

Conclusion

Permaculture gardening offers a comprehensive and impactful approach to environmental sustainability. By focusing on soil health, water conservation, biodiversity, and climate resilience, permaculture gardens create self-sustaining ecosystems that benefit both people and the planet. These practices reduce pollution, enhance local food systems, and promote regenerative agriculture, making permaculture a powerful tool for addressing the environmental challenges of our time.

As the global community seeks solutions to climate change, food insecurity, and environmental degradation, permaculture gardening provides a hopeful and practical path forward. Its principles and practices empower individuals and communities to live sustainably, regenerate natural systems, and build a resilient future.

Personal Benefits

Personal Benefits of Permaculture Gardening

Permaculture gardening is not just a method for sustainable living and environmental stewardship; it also offers numerous personal benefits. Engaging in permaculture can transform individuals' lives by enhancing physical health, mental well-being, and social connections. This approach promotes a deeper understanding of nature, encourages self-sufficiency, and fosters a sense of purpose and accomplishment. In this comprehensive exploration, we will delve into the various personal benefits of permaculture gardening.

Physical Health Benefits

1. Improved Nutrition:
   - Growing your own food through permaculture gardening ensures access to fresh, organic, and nutrient-dense produce. Consuming fruits and vegetables straight from the garden provides higher levels of vitamins, minerals, and antioxidants compared to store-bought counterparts, which often lose nutrients during transportation and storage.
   - A diverse permaculture garden can offer a wide variety of crops, promoting a balanced and varied diet. Eating seasonally

from your garden aligns with natural cycles and supports optimal health.

2. Increased Physical Activity:
   - Gardening is a form of moderate physical exercise that involves activities such as digging, planting, weeding, and harvesting. These activities improve cardiovascular health, strength, flexibility, and endurance.
   - Regular physical activity through gardening can help maintain a healthy weight, reduce the risk of chronic diseases such as heart disease, diabetes, and osteoporosis, and improve overall physical fitness.

3. Exposure to Fresh Air and Sunlight:
   - Spending time outdoors in the garden provides exposure to fresh air and sunlight, both of which are beneficial for health. Fresh air can enhance respiratory function, while sunlight helps the body produce vitamin D,

which is crucial for bone health, immune function, and mood regulation.

- Gardening in natural light can help regulate circadian rhythms, improving sleep patterns and overall energy levels.

Mental Health and Emotional Well-Being

1. Stress Reduction and Relaxation:

- Gardening has been shown to reduce stress and promote relaxation. The act of working with plants, soil, and nature can be meditative, providing a sense of calm and tranquility.

- Engaging in repetitive gardening tasks can help lower cortisol levels (a stress hormone) and promote the release of endorphins, which are natural mood enhancers.

2. Mental Clarity and Cognitive Function:

- The sensory experiences of gardening—such as the sights, sounds, and smells of nature—can stimulate the mind and improve cognitive function. Studies have shown that spending time in nature can enhance attention, memory, and creativity.

- Problem-solving and planning involved in designing and maintaining a permaculture garden can keep the mind active and sharp, reducing the risk of cognitive decline with age.

3. Sense of Accomplishment and Purpose:

   - Successfully growing and harvesting food from a permaculture garden provides a sense of accomplishment and purpose. Witnessing the fruits of your labor and the tangible results of your efforts can boost self-esteem and confidence.

   - Gardening projects offer goals to work towards and a sense of progression, which can be particularly beneficial for individuals experiencing feelings of stagnation or lack of direction.

Social and Community Benefits

1. Strengthened Social Connections:
- Permaculture gardening often involves community participation and collaboration. Community gardens, urban farms, and cooperative projects provide opportunities to connect with others, fostering social interactions and building relationships.
- Shared gardening experiences can create a sense of community and belonging, reducing feelings of isolation and loneliness.

2. Opportunities for Education and Sharing Knowledge:
- Permaculture gardening offers opportunities to learn and share knowledge about sustainable practices, ecological principles, and food production. Teaching and learning from others can be enriching and fulfilling.
- Community workshops, garden tours, and educational programs can enhance

social bonds and create a culture of mutual support and learning.

3. Intergenerational Engagement:
   - Gardening can bring together people of all ages, from children to seniors. Intergenerational engagement through gardening activities promotes the exchange of wisdom and experience, strengthening family and community ties.
   - Involving children in gardening helps them develop a connection to nature, an understanding of where food comes from, and valuable life skills.

Self-Sufficiency and Empowerment

1. Food Security and Independence:
   - Growing your own food through permaculture gardening enhances food security and reduces reliance on external food sources. This independence can be particularly empowering during times of

economic instability or supply chain disruptions.

- Producing a portion of your own food can reduce grocery bills and provide a buffer against rising food prices, contributing to financial stability.

2. Resourcefulness and Problem-Solving Skills:

- Permaculture gardening encourages resourcefulness and creativity. Finding solutions to gardening challenges, such as pest control, soil improvement, and water management, fosters problem-solving skills and adaptability.

- Repurposing materials, creating homemade compost, and designing efficient garden systems enhance self-reliance and ingenuity.

3. Sustainable Living and Ethical Responsibility:

- Practicing permaculture gardening instills a sense of ethical responsibility and stewardship for the environment. By adopting sustainable practices, individuals contribute to the well-being of the planet and future generations.

CHAPTER 1

Getting Started

Getting Started with Permaculture Gardening

Embarking on a permaculture gardening journey involves understanding the core principles of permaculture, planning your garden design, and implementing sustainable practices. This comprehensive guide will help you get started with permaculture gardening, providing you with the essential steps and considerations for creating a productive and sustainable garden.

Step 1: Understanding Permaculture Principles

Permaculture is based on a set of principles that guide the design and implementation of sustainable systems. Familiarize yourself with these principles to inform your gardening practices:

1. Observe and Interact:
   - Spend time observing your garden site to understand its unique characteristics, such as sunlight, wind patterns, soil type, and existing vegetation.
      - Interact with your environment and consider how different elements (plants,

animals, structures) can work together harmoniously.

2. Catch and Store Energy:
- Use resources efficiently by capturing and storing energy. For example, collect rainwater, harness solar energy, and build healthy soil that retains nutrients and moisture.

3. Obtain a Yield:
- Ensure that your garden provides tangible benefits, such as food, medicine, and other resources. Focus on productive plants and sustainable harvests.

4. Apply Self-Regulation and Accept Feedback:
- Monitor your garden's performance and make adjustments as needed. Learn from successes and failures to improve your system.

5. Use and Value Renewable Resources and Services:

- Prioritize renewable resources over non-renewable ones. For instance, use compost instead of synthetic fertilizers and plant perennials alongside annuals.

6. Produce No Waste:

- Design systems that minimize waste by recycling and reusing materials. Compost organic waste, use greywater for irrigation, and repurpose garden materials.

7. Design from Patterns to Details:

- Start with a broad vision for your garden and refine the details as you go. Look at natural patterns and mimic them in your garden design.

8. Integrate Rather Than Segregate:

- Create connections between different elements in your garden. Companion planting, guilds, and polycultures can enhance productivity and resilience.

9. Use Small and Slow Solutions:
   - Implement gradual changes and manageable projects. Small, incremental steps allow you to adapt and refine your approach.

10. Use and Value Diversity:
   - Incorporate a wide variety of plants, animals, and microorganisms to create a resilient and productive garden.

11. Use Edges and Value the Marginal:
   - Utilize the edges and transitional areas in your garden, where different elements meet. These areas often have high productivity and biodiversity.

12. Creatively Use and Respond to Change:
   - Embrace change and adapt your garden to evolving conditions. Be flexible and innovative in your approach.

Step 2: Site Assessment and Planning

1. Site Analysis:
   - Conduct a thorough site analysis to understand the unique characteristics of your garden space. Consider factors such as sunlight, shade, wind, slope, soil type, and existing vegetation.
   - Map out your garden area, noting key features and potential challenges.

2. Define Your Goals:
   - Clarify your goals for the garden. What do you want to achieve? Consider factors such as food production, biodiversity, aesthetics, and educational opportunities.

- Identify your priorities and focus areas, such as growing vegetables, creating a wildlife habitat, or building soil health.

3. Design Your Garden Layout:
 - Create a garden design that incorporates permaculture principles. Consider zoning your garden based on frequency of use, with Zone 1 being the most frequently visited and Zone 5 the least.
 - Plan for key elements such as garden beds, paths, water features, composting areas, and structures (e.g., trellises, greenhouses).

Step 3: Building Healthy Soil

1. Soil Testing:
 - Test your soil to determine its pH, nutrient levels, and texture. This information will help you make informed decisions about soil amendments and planting.
 - Use a home testing kit or send a sample to a laboratory for analysis.

2. Improving Soil Fertility:

- Add organic matter to your soil through composting, mulching, and cover cropping. Organic matter improves soil structure, water retention, and nutrient availability.
- Use green manures and cover crops to build soil fertility and prevent erosion.

3. Implementing No-Till Practices:

- Avoid tilling the soil, which can disrupt soil structure and microbial life. Instead, use no-till practices such as sheet mulching or lasagna gardening to build healthy soil layers.
- Mulch with organic materials such as straw, leaves, or wood chips to suppress weeds, retain moisture, and add nutrients to the soil.

Step 4: Water Management

1. Rainwater Harvesting:

- Install rain barrels or cisterns to collect and store rainwater for garden use. This reduces your reliance on municipal water and provides a sustainable water source.

- Ensure proper filtration and storage to maintain water quality.

2. Irrigation Techniques:

- Use efficient irrigation methods such as drip irrigation or soaker hoses to deliver water directly to the root zone of plants, minimizing evaporation and runoff.

- Water in the early morning or late evening to reduce water loss due to evaporation.

3. Swales and Contour Planting:

- Create swales (shallow trenches) along the contour of your garden to capture and infiltrate rainwater. Swales help to slow down water flow, reduce erosion, and improve soil moisture.

- Plant along the contour lines to take advantage of natural water movement and improve water distribution.

Step 5: Plant Selection and Diversity

1. Choosing the Right Plants:
   - Select plants that are well-suited to your local climate, soil, and growing conditions. Consider factors such as hardiness, water needs, and pest resistance.
   - Include a mix of annuals, perennials, vegetables, herbs, flowers, and native plants to enhance biodiversity and resilience.

2. Companion Planting and Guilds:
   - Use companion planting to promote beneficial interactions between plants. For example, plant basil near tomatoes to repel pests and improve flavor.
   - Create plant guilds, which are groups of plants that support each other's growth. A typical guild might include a fruit tree, nitrogen-fixing plants, ground covers, and pollinator-attracting flowers.

3. Succession Planting:

- Plan for succession planting to ensure continuous harvests throughout the growing season. Replace harvested crops with new plantings to maximize productivity.
- Use intercropping and relay planting to make efficient use of space and resources.

Step 6: Natural Pest and Weed Management

1. Attracting Beneficial Insects:

- Plant a variety of flowers and herbs that attract beneficial insects such as ladybugs, bees, and predatory wasps. These insects help control pests naturally.
- Provide habitat features such as insect hotels, water sources, and sheltered areas to support beneficial insect populations.

2. Integrated Pest Management (IPM):

- Implement an integrated pest management approach that combines cultural, biological, and mechanical control methods. Monitor pest populations and take action only when necessary.

- Use barriers, traps, and natural repellents to manage pests without harmful chemicals.

3. Weed Control:
- Mulch garden beds to suppress weeds and retain soil moisture. Organic mulches such as straw, wood chips, and leaves are effective and add nutrients to the soil as they decompose.

- Use hand weeding, hoeing, and other manual methods to control weeds. Regular maintenance prevents weeds from becoming established and competing with your crops.

Step 7: Continuous Learning and Adaptation

1. Observation and Adjustment:
- Regularly observe your garden to assess its health, productivity, and resilience. Take

note of what works well and what needs improvement.

- Be willing to adjust your practices based on feedback from your garden. Permaculture is an iterative process that involves continuous learning and adaptation.

2. Educational Resources:

- Take advantage of educational resources such as books, online courses, workshops, and community groups to deepen your knowledge of permaculture principles and practices.

- Engage with the permaculture community to share experiences, seek advice, and stay inspired.

3. Documenting Your Journey:

- Keep a garden journal to record your observations, successes, challenges, and experiments. Documenting your journey helps you track progress, learn from experiences, and celebrate achievements.

- Use photos, sketches, and notes to create a visual and written record of your garden's evolution.

Conclusion

Getting started with permaculture gardening involves understanding its core principles, conducting a thorough site assessment, and implementing sustainable practices. By focusing on soil health, water management, plant diversity, and natural pest control, you can create a productive and resilient garden. Permaculture gardening is a rewarding journey that offers numerous personal benefits, from improved physical health to enhanced mental well-being and a deeper connection to nature. Embrace the principles of permaculture, continue learning and adapting, and enjoy the process of creating a sustainable and thriving garden.

1. Understanding Your Space

Before embarking on your permaculture gardening journey, it's crucial to thoroughly understand the space you will be working with. A detailed site assessment and analysis will inform your design decisions and help you create a productive, sustainable, and resilient garden. This comprehensive guide will walk you through the essential steps and considerations for understanding your space in the context of permaculture gardening.

Step 1: Conducting a Site Assessment

1. Observing and Recording:
   - Spend time observing your garden site at different times of the day and throughout the seasons. Take note of how the space changes with varying weather conditions, light levels, and temperatures.

- Keep a detailed journal or sketchbook to record your observations, including sketches, notes, and photographs.

2. Understanding Climate and Microclimates:

- Gather information about your region's climate, including average temperatures, rainfall patterns, frost dates, and prevailing winds. This data will help you select appropriate plants and design features.
- Identify microclimates within your garden space. Microclimates are areas with distinct environmental conditions, such as sunny spots, shady corners, or areas sheltered from the wind. Understanding microclimates allows you to optimize plant placement and garden design.

3. Mapping Sunlight and Shade:

- Track the path of the sun across your garden to understand where and when different areas receive sunlight and shade.

Use a sun chart or sun path diagram to map these patterns.

- Identify areas that receive full sun, partial shade, or full shade. This information will guide your plant selection and placement, ensuring that each plant receives the appropriate amount of sunlight.

4. Analyzing Soil:

- Conduct a soil test to determine the soil type, pH, nutrient levels, and organic matter content. You can use a home testing kit or send a soil sample to a laboratory for analysis.
- Assess soil texture by feeling it between your fingers. Soil texture (sand, silt, clay) affects water retention, drainage, and root growth. Understanding your soil type will help you amend and manage it effectively.

5. Identifying Water Sources and Drainage Patterns:

- Identify existing water sources such as rainwater, greywater, and municipal water.

Consider how you can harvest and store rainwater for irrigation.

- Observe drainage patterns to identify areas prone to waterlogging or erosion. Plan for swales, rain gardens, or other water management features to address these issues.

6. Evaluating Existing Vegetation:
- Take an inventory of existing plants, trees, and shrubs in your garden. Note their health, growth patterns, and interactions with other elements.

- Assess whether existing vegetation supports your permaculture goals. You may need to remove invasive species, prune overgrown plants, or integrate new plants that enhance biodiversity and productivity.

7. Considering Wildlife and Ecosystems:
- Observe the presence of wildlife in your garden, including birds, insects, mammals, and reptiles. Identify beneficial species that

can support your garden ecosystem, as well as potential pests.

- Consider how your garden can provide habitat and resources for wildlife, such as nesting sites, food sources, and water features.

Step 2: Mapping and Zoning Your Garden

1. Creating a Base Map:
- Draw a base map of your garden, including boundaries, structures, pathways, water sources, and existing vegetation. Use graph paper or digital mapping tools to create an accurate and scaled representation.
- Include significant features such as slopes, fences, gates, and utilities. A detailed base map provides a foundation for your garden design.

2. Zoning Your Garden:

- Divide your garden into zones based on frequency of use and accessibility. Permaculture typically uses five zones, with Zone 1 being the most frequently visited and Zone 5 the least:

   - **Zone 1:** Areas close to your home, including kitchen gardens, herb beds, and daily-use areas.
   - **Zone 2:** Areas used less frequently, such as vegetable patches, compost bins, and small livestock enclosures.
   - **Zone 3:** Areas for larger-scale production, such as orchards, fields, and larger animal pastures.
   - **Zone 4:** Semi-wild areas for foraging, timber, and wildlife habitat.
   - **Zone 5:** Untouched wilderness areas for observation and conservation.

- Plan the placement of elements within each zone to maximize efficiency and minimize effort. For example, place frequently used plants and tools in Zone 1 for easy access.

3. Sector Analysis:
- Conduct a sector analysis to identify external factors that affect your garden, such as wind direction, sun angles, noise sources, and potential hazards. Mark these sectors on your base map.
- Use sector analysis to inform design decisions, such as windbreaks, sun traps, noise barriers, and buffer zones. This helps you create a garden that responds to and leverages external influences.

Step 3: Designing Your Permaculture Garden

1. Designing for Energy Efficiency:
- Plan your garden layout to optimize energy use, both human and natural. Place elements that require frequent attention or produce high yields (such as vegetable beds) close to your home.
- Use natural elements such as trees, hedges, and water features to create microclimates, improve energy efficiency,

and enhance the overall garden environment.

2. Incorporating Key Permaculture Elements:

- Integrate key permaculture elements into your garden design, including:
    - **Water management**: Swales, ponds, rain gardens, and irrigation systems.
    - **Soil health:** Compost bins, worm farms, and mulching systems.
    - **Biodiversity:** Diverse plantings, polycultures, and habitat features.
    - **Renewable resources:** Solar panels, rainwater harvesting systems, and natural building materials.
- Ensure that each element serves multiple functions and supports the overall garden ecosystem.

3. Implementing Plant Guilds and Companion Planting:

- Design plant guilds, which are groups of plants that support each other's growth and create a balanced ecosystem. A typical guild might include a central tree, nitrogen-fixing plants, ground covers, and pollinator-attracting flowers.

- Use companion planting to enhance plant health, deter pests, and improve yields. Research beneficial plant combinations and incorporate them into your garden design.

4. Planning for Succession Planting:

- Plan for succession planting to ensure continuous harvests throughout the growing season. Replace harvested crops with new plantings to maximize productivity and maintain soil health.

- Use intercropping and relay planting to make efficient use of space and resources, ensuring a steady supply of fresh produce.

Step 4: Implementing Your Garden Design

1. Starting Small and Scaling Up:

- Begin with small, manageable projects to gain experience and confidence. Focus on key areas such as a kitchen garden or herb bed before expanding to larger zones.

- Gradually scale up your efforts as you learn and adapt, building a resilient and productive garden over time.

2. Building Healthy Soil:

- Implement soil-building practices such as composting, mulching, and cover cropping to enhance soil fertility and structure.

- Use no-till methods to protect soil integrity and promote beneficial soil microorganisms.

3. Installing Water Management Systems:

- Set up rainwater harvesting systems, such as rain barrels or cisterns, to collect and store rainwater for irrigation.

- Construct swales, ponds, and other water management features to capture and manage water on-site, reducing runoff and erosion.

4. Planting and Mulching:
 - Choose a diverse range of plants suited to your local climate and soil conditions. Include annuals, perennials, vegetables, herbs, and native plants to enhance biodiversity.
 - Mulch garden beds with organic materials to retain moisture, suppress weeds, and improve soil health.

5. Creating Habitat and Encouraging Wildlife:
 - Install habitat features such as birdhouses, insect hotels, and ponds to attract and support beneficial wildlife.
 - Plant a variety of flowers and herbs to attract pollinators and beneficial insects, enhancing garden health and productivity.

Step 5: Monitoring, Learning, and Adapting

1. Regular Observation and Documentation:

- Continuously observe your garden to assess its health, productivity, and resilience. Take note of successes, challenges, and changes over time.
- Keep a garden journal to document your observations, experiments, and progress. This record will help you learn and adapt your practices.

2. Learning from Experience:
- Reflect on your experiences and adjust your practices based on feedback from your garden. Permaculture is an iterative process that involves continuous learning and improvement.
- Seek out educational resources, such as books, courses, workshops, and community groups, to deepen your knowledge and stay inspired.

3. Sharing Knowledge and Building Community:
- Share your experiences and knowledge with others, contributing to a broader culture of sustainability and resilience.

- Engage with the permaculture community to exchange ideas, seek advice, and collaborate on projects, fostering a supportive and connected network.

Conclusion

Understanding your space is a crucial first step in creating a successful permaculture garden. By conducting a thorough site assessment, mapping and zoning your garden, and designing with permaculture principles in mind, you can create a productive, sustainable, and resilient garden. Implementing key elements, building healthy soil, and managing water effectively will set the foundation for a thriving garden ecosystem. Regular observation, learning, and adaptation will help you refine your practices and achieve your permaculture goals. Embrace the journey, enjoy the process, and reap the rewards of a well-designed permaculture garden.

Observing and Assessing Your Garden Area

Observing and assessing your garden area is the foundation of successful permaculture design. This process involves understanding the unique characteristics of your site, such as sunlight, soil, water, climate, and existing vegetation. By thoroughly evaluating these factors, you can make informed decisions that align with permaculture principles and create a productive and sustainable garden. This guide provides an extensive and comprehensive approach to observing and assessing your garden area.

Step 1: Initial Observation and Recording

1. Spend Time in Your Garden:
   - Spend several hours, days, and even seasons in your garden to observe its dynamics. Note how the environment changes throughout the day and year.

- Pay attention to the natural patterns and processes, such as how sunlight moves, where water collects, and which areas are sheltered from the wind.

2. Create a Garden Journal:
- Keep a detailed journal or sketchbook to record your observations. Include sketches, notes, photographs, and maps.
- Document key features, such as existing plants, structures, soil conditions, and wildlife. Record any changes or patterns you notice over time.

- **Step 2: Analyzing Sunlight and Shade**

- **1. Track the Sun's Path:**
  - Observe the sun's path across your garden at different times of the day and throughout the seasons. Use tools like a sun chart or sun path diagram to map these patterns.

- Identify areas that receive full sun, partial shade, or full shade. Note any changes in light conditions due to seasonal variations.

2. Shade Analysis:
- Identify sources of shade, such as trees, buildings, fences, and other structures. Determine how these elements affect light availability in different parts of your garden.
- Consider the impact of shade on plant growth and select plants that thrive in the available light conditions.

3. Solar Access for Structures:
- If you plan to install structures like greenhouses, cold frames, or solar panels, ensure they receive adequate sunlight. Place them in locations that maximize solar access while considering seasonal variations.

Step 3: Soil Assessment

1. Soil Testing:
 - Conduct a soil test to determine the soil's pH, nutrient levels, and organic matter content. Use a home testing kit or send a soil sample to a laboratory for analysis.
 - Test for contaminants, especially if your garden is in an urban area or near industrial sites. Soil contamination can affect plant health and safety.

2. Soil Texture and Structure:
 - Assess soil texture by feeling it between your fingers. Determine if it's sandy, silty, clayey, or loamy. Soil texture affects water retention, drainage, and root growth.
 - Evaluate soil structure by digging a small hole and examining the soil profile. Healthy soil has a crumbly texture with visible organic matter and earthworms.

3. Drainage and Moisture Levels:

- Observe how water moves through your garden during and after rain. Identify areas with poor drainage, standing water, or erosion.

- Test soil drainage by digging a small hole, filling it with water, and measuring how long it takes to drain. Well-drained soil should absorb water within a few hours.

Step 4: Water Sources and Management

1. Identify Water Sources:

- Determine the sources of water available for your garden, such as rainwater, greywater, municipal water, and natural water bodies. Consider how you can harvest and store rainwater for irrigation.

- Assess the reliability and sustainability of these water sources, especially during dry seasons or droughts.

2. Map Drainage Patterns:

- Observe how water flows through your garden during rainfall. Identify areas prone to waterlogging, erosion, or runoff.

- Use contour lines and swales to capture and redirect water, improving soil moisture and preventing erosion.

3. Water Conservation Techniques:

- Implement water-saving techniques such as mulching, drip irrigation, and rainwater harvesting. Mulching helps retain soil moisture, reduce evaporation, and suppress weeds.

- Install rain barrels or cisterns to collect and store rainwater for garden use. Ensure proper filtration and storage to maintain water quality.

Step 5: Climate and Microclimates

1. Understand Regional Climate:

- Research your region's climate, including average temperatures, rainfall patterns, frost dates, and prevailing winds. This information will guide plant selection and garden design.

- Use climate data to plan for seasonal variations, such as planting dates, frost protection, and heat mitigation.

2. Identify Microclimates:

- Identify microclimates within your garden, which are areas with distinct environmental conditions. Microclimates can be created by factors such as sunlight, shade, wind, and topography.

- Use microclimates to your advantage by selecting plants that thrive in specific conditions. For example, plant heat-loving crops in sunny, sheltered spots and shade-tolerant plants in cooler areas.

3. Wind Patterns and Protection:

- Observe the direction and intensity of prevailing winds. Strong winds can damage plants, dry out soil, and affect pollination.

- Use windbreaks such as hedges, trees, or fences to protect your garden from harsh winds. Design windbreaks to create sheltered microclimates and reduce wind speed.

Step 6: Existing Vegetation and Wildlife

1. Inventory Existing Plants:
   - Take an inventory of existing plants, trees, and shrubs in your garden. Note their health, growth patterns, and interactions with other elements.
   - Assess whether existing vegetation supports your permaculture goals. You may need to remove invasive species, prune overgrown plants, or integrate new plants to enhance biodiversity and productivity.

2. Assessing Plant Health:
   - Examine the health of existing plants. Look for signs of disease, pests, nutrient

deficiencies, or stress. Healthy plants indicate suitable growing conditions, while unhealthy plants may signal underlying issues.

- Determine the age and lifespan of existing plants. Consider whether they need replacement, rejuvenation, or support to thrive.

3. Observing Wildlife:

- Observe the presence of wildlife in your garden, including birds, insects, mammals, and reptiles. Identify beneficial species that can support your garden ecosystem, as well as potential pests.

- Create habitats and resources for beneficial wildlife, such as birdhouses, insect hotels, and water features. Encourage biodiversity by planting a variety of flowers and herbs to attract pollinators and beneficial insects.

Step 7: Identifying and Addressing Potential Hazards

1. Potential Environmental Hazards:

- Identify environmental hazards such as flooding, landslides, soil contamination, and pollution. Address these issues through appropriate design and management strategies.
- Consider the impact of extreme weather events, such as storms, droughts, and heatwaves. Plan for resilience and mitigation measures to protect your garden.

2. Human-made Hazards:

- Assess potential human-made hazards, such as pesticide drift, noise pollution, and encroaching development. Implement buffer zones, barriers, and protective measures to mitigate these risks.
- Engage with neighbors and local authorities to address community-wide environmental concerns and promote sustainable practices.

3. Safety and Accessibility:

- Ensure your garden design prioritizes safety and accessibility. Plan for safe pathways, stable structures, and easy access to key areas.

- Consider the needs of all garden users, including children, elderly individuals, and people with disabilities. Design inclusive and accessible spaces that everyone can enjoy.

Step 8: Using Mapping and Digital Tools

1. Creating a Base Map:

- Draw a base map of your garden, including boundaries, structures, pathways, water sources, and existing vegetation. Use graph paper or digital mapping tools to create an accurate and scaled representation.

- Include significant features such as slopes, fences, gates, and utilities. A detailed base map provides a foundation for your garden design.

2. Using Digital Tools and Apps:

- Utilize digital tools and apps for mapping, design, and data collection. Tools such as Google Earth, GIS software, and garden planning apps can enhance your site assessment and design process.

- Use technology to gather and analyze data, track progress, and visualize your garden design. Digital tools can streamline planning, communication, and documentation.

3. Incorporating Layers and Overlays:

- Create layered maps and overlays to visualize different aspects of your garden, such as sunlight, soil, water, and vegetation. Layers help you analyze interactions and relationships between elements.

- Use overlays to identify areas for improvement, plan for future projects, and monitor changes over time. A multi-layered approach provides a comprehensive understanding of your garden.

Conclusion

Observing and assessing your garden area is a critical step in creating a successful permaculture design. By thoroughly evaluating factors such as sunlight, soil, water, climate, existing vegetation, and potential hazards, you can make informed decisions that align with permaculture principles. Use detailed observations, mapping, and digital tools to gather and analyze data, guiding your design and implementation process. A comprehensive understanding of your garden area sets the foundation for a productive, sustainable, and resilient permaculture garden. Embrace the process, learn from your observations, and adapt your practices to create a thriving garden ecosystem.

Climate and Microclimates

Climate and Microclimates: A Comprehensive Guide for Permaculture Gardening

Understanding the climate and microclimates of your garden is essential for successful permaculture design. Climate influences the types of plants you can grow, while microclimates allow for the creation of diverse growing conditions within a single garden. This guide provides an in-depth look at both broader climatic factors and the specific microclimates within your garden, offering strategies to optimize these conditions for a productive and resilient permaculture system.

Understanding Climate

1. Defining Climate:
   - Climate refers to the long-term patterns of temperature, humidity, wind, and precipitation in a particular region. It differs from weather, which describes short-term atmospheric conditions.

- Climate influences the growing season, plant hardiness, and overall garden productivity.

2. Types of Climate:
   - **Tropical**: Warm temperatures year-round with significant rainfall. Supports a wide variety of plants, including tropical fruits and perennial crops.
   - **Temperate:** Distinct seasons with moderate temperatures. Suitable for a diverse range of annual and perennial plants.
   - **Arid:** Hot, dry conditions with limited rainfall. Requires drought-tolerant plants and efficient water management.
   - **Mediterranean**: Mild, wet winters and hot, dry summers. Supports a variety of fruits, vegetables, and herbs adapted to seasonal variations.
   - **Cold:** Long, harsh winters with short growing seasons. Suitable for cold-hardy plants and season extension techniques.

3. Key Climatic Factors:

- **Temperature:** Influences plant growth rates, flowering, and fruiting. Extreme temperatures can stress plants and reduce yields.
- **Precipitation:** Determines water availability for plants. Both excessive and insufficient rainfall can pose challenges.
- **Humidity:** Affects transpiration rates, disease prevalence, and plant stress. High humidity can increase fungal diseases, while low humidity can cause drought stress.
- **Wind:** Impacts pollination, plant structure, and water loss. Strong winds can damage plants and reduce yields.
- **Seasonality:** Includes variations in day length, temperature, and weather patterns throughout the year. Understanding seasonality helps in planning planting schedules and crop rotations.

4. Regional Climate Data:

- Obtain regional climate data from local weather stations, agricultural extensions, or online databases. This information helps in selecting suitable plants and planning garden activities.

- Use climate data to determine average frost dates, rainfall patterns, temperature ranges, and wind speeds.

Identifying Microclimates

1. Defining Microclimates:
- Microclimates are localized areas within a garden that have distinct environmental conditions. These can be influenced by factors such as topography, vegetation, water bodies, and man-made structures.
- Microclimates can create unique growing conditions, allowing for greater plant diversity and productivity.

2. Factors Influencing Microclimates:
- **Topography**: Slopes, valleys, and elevation changes can create variations in sunlight, temperature, and moisture.

South-facing slopes in the Northern Hemisphere receive more sunlight and warmth, while north-facing slopes remain cooler and shadier.

- **Vegetation:** Trees, shrubs, and ground cover can modify microclimates by providing shade, wind protection, and moisture retention. Dense vegetation can create cooler, more humid conditions.

- **Water Bodies:** Lakes, ponds, and streams can moderate temperature extremes and increase humidity. Proximity to water can create milder microclimates that extend growing seasons.

- **Structures:** Buildings, walls, fences, and other structures can block wind, create shade, and reflect sunlight. These features can create warmer or cooler spots within the garden.

- **Soil:** Variations in soil type, texture, and moisture content can influence microclimates. Well-drained, sandy soils may be warmer and drier, while clay soils retain more moisture and remain cooler.

3. Creating and Modifying Microclimates:

- Use garden design elements to create or enhance microclimates. For example, plant windbreaks to reduce wind speed and create sheltered areas, or install reflective surfaces to increase sunlight and warmth.

- Modify existing microclimates by adjusting irrigation, mulching, or soil amendments. For instance, mulching can help retain soil moisture and create a cooler microclimate for shade-loving plants.

Analyzing Your Garden's Climate and Microclimates

1. Mapping and Recording:

- Create a detailed map of your garden, marking areas with different microclimates. Use tools like sun charts, soil thermometers, and moisture meters to gather data.

- Record observations over time to understand how microclimates change with the seasons. Note areas that receive more or less sunlight, retain moisture, or are exposed to wind.

2. Using Technology:
- Utilize digital tools and apps to analyze and visualize microclimates. Tools like GIS software, garden planning apps, and climate data models can provide valuable insights.
- Use sensors and monitoring devices to track real-time environmental conditions, such as soil moisture, temperature, and humidity. This data helps in making informed decisions about plant placement and garden management.

3. Integrating Observations:
- Combine your observations with regional climate data to create a comprehensive understanding of your garden's environmental conditions. This integrated approach helps in selecting suitable plants and designing resilient garden systems.

- Share your findings with other gardeners and permaculture enthusiasts to gain insights and learn from their experiences.

Strategies for Working with Climate and Microclimates

1. Selecting Appropriate Plants:
   - Choose plants that are well-suited to your regional climate and specific microclimates. Consider factors such as temperature tolerance, water needs, and sunlight requirements.
   - Use plant hardiness zones as a guide for selecting perennials and trees. Choose annuals and vegetables that match your growing season length and climate conditions.

2. Designing for Climate Resilience:
   - Incorporate design elements that enhance climate resilience, such as windbreaks, shade structures, and rain gardens. These features help mitigate the

impacts of extreme weather events and create more stable growing conditions.
 - Plan for seasonal variations by using techniques like succession planting, crop rotation, and intercropping. These practices help maintain soil health, optimize resource use, and extend the growing season.

3. Microclimate-Specific Planting:
 - Place plants in locations that match their microclimate preferences. For example, plant heat-loving crops in sunny, south-facing areas, and shade-tolerant plants in cooler, north-facing spots.
 - Use companion planting and plant guilds to create mutually beneficial relationships between plants. For instance, plant tall, sun-loving crops to provide shade for shorter, shade-tolerant plants.

4. Water Management:
 - Implement water management strategies that suit your climate and microclimates.

Use techniques like mulching, drip irrigation, and rainwater harvesting to conserve water and maintain soil moisture.

- Design your garden to capture and retain water, using features like swales, berms, and retention ponds. These elements help manage water flow, reduce erosion, and create moist microclimates.

5. Soil Management:

- Enhance soil health and structure through practices like composting, cover cropping, and no-till gardening. Healthy soil supports resilient plants and helps buffer against climate extremes.

- Use organic matter and mulch to improve soil moisture retention and create favorable microclimates for plant roots. These practices help maintain consistent growing conditions and reduce stress on plants.

Conclusion

Understanding and working with the climate and microclimates of your garden is essential for successful permaculture gardening. By thoroughly analyzing regional climate data and observing microclimates within your garden, you can make informed decisions that optimize growing conditions. Utilize garden design elements, plant selection, and management practices to create a productive and resilient garden that thrives in its unique environment. Embrace the diversity of microclimates within your garden, and leverage them to enhance biodiversity, productivity, and sustainability.

2. Design Principles

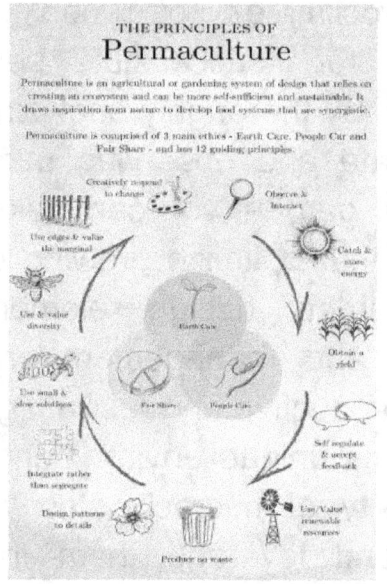

Permaculture Design Principles: An Extensive and Comprehensive Guide

Permaculture design principles form the foundation of creating sustainable, resilient, and productive systems. Developed by Bill Mollison and David Holmgren, these principles guide the way we approach and interact with natural systems to create harmonious and regenerative environments. This guide delves deeply into each of the key permaculture design principles, offering a thorough understanding of their application and importance in permaculture gardening.

1. Observe and Interact

Principle Overview:
- Observation is the cornerstone of permaculture design. By taking the time to watch and understand natural systems, you can make informed decisions that work with nature rather than against it.
- Interaction follows observation; it involves actively engaging with your environment to test, adapt, and refine your strategies.

Application:
- Spend time observing your garden across different seasons and weather conditions. Note sunlight patterns, wind direction, water flow, and existing plant and animal life.

- Use a garden journal to record your observations and track changes over time.
- Engage with your garden through small, incremental experiments. Plant a variety of crops and observe which thrive best in your conditions.

2. Catch and Store Energy

Principle Overview:
- Energy, in various forms, flows through natural systems. Capturing and storing this energy ensures it is available when needed, enhancing resilience and productivity.
- This principle includes solar energy, water, wind, biomass, and human energy.

Application:
- **Solar Energy:** Install solar panels or use passive solar design to harness sunlight for electricity and heating. Position your garden to maximize sunlight exposure for photosynthesis.

- **Water**: Capture rainwater using barrels, ponds, or swales. Store it for irrigation during dry periods.
- **Biomass:** Grow and compost organic matter to create a sustainable source of nutrients for your soil.
- **Wind:** Utilize windbreaks and small wind turbines to protect plants and generate power.

3. Obtain a Yield

Principle Overview:
- Every element in your permaculture design should produce a tangible benefit or yield. This ensures that your efforts contribute to your well-being and the health of your ecosystem.
- Yields can be food, fiber, fuel, medicine, or other resources that support human and ecological health.

Application:

- Design your garden to produce a diverse range of yields. Grow a variety of fruits, vegetables, herbs, and medicinal plants.
- Incorporate animals such as chickens or bees to obtain yields of eggs, honey, and pollination services.
- Harvest rainwater, compost, and biomass to create yields of water and soil fertility.

4. Apply Self-Regulation and Accept Feedback

Principle Overview:
- Self-regulation involves setting limits and creating systems that can manage themselves with minimal intervention.
- Accepting feedback means learning from your mistakes and successes, and adjusting your practices accordingly.

Application:
- Monitor your garden for signs of imbalance or inefficiency. This could include pest

outbreaks, nutrient deficiencies, or overuse of resources.
- Use integrated pest management (IPM) to regulate pests naturally, reducing the need for chemical interventions.
- Adjust planting schedules, water usage, and soil management practices based on feedback from your garden.

5. Use and Value Renewable Resources and Services

Principle Overview:
- Prioritize renewable resources and ecosystem services that regenerate over time, reducing reliance on non-renewable resources.
- This principle emphasizes sustainability and the conservation of energy and materials.

Application:
- Use renewable energy sources such as solar, wind, and hydro power for your garden needs.
- Opt for hand tools and natural materials over synthetic and fossil-fuel-based products.
- Plant perennials and nitrogen-fixing plants to provide ongoing benefits to your soil and ecosystem.

6. Produce No Waste

Principle Overview:
- In nature, waste is a resource. Applying this principle means designing systems that repurpose all outputs into inputs for other processes.
- This principle encourages recycling, composting, and mindful consumption.

Application:
- Implement a comprehensive composting system to recycle kitchen and garden waste into valuable soil amendments.
- Use greywater systems to recycle household water for irrigation.
- Repurposed materials like wood, metal, and plastic for garden structures and tools.

7. Design from Patterns to Details

Principle Overview:
- Start with the big picture – the natural patterns and flows in your environment – and then refine your design to the finer details.
- Recognizing and mimicking natural patterns ensures that your design works harmoniously with nature.

Application:
- Identify and map patterns such as wind direction, water flow, and sunlight exposure in your garden.

- Design broad zones and sectors based on these patterns, then refine the layout of garden beds, paths, and structures.
- Use natural shapes like spirals, circles, and contours in your garden design to enhance efficiency and aesthetics.

8. Integrate Rather Than Segregate

Principle Overview:
- Successful permaculture systems feature elements that work together synergistically. Integration maximizes cooperation and minimizes competition.
- The principle promotes diversity and interdependence, creating resilient and productive ecosystems.

Application:
- Practice companion planting to enhance plant growth, repel pests, and improve soil health.

- Create plant guilds, where groups of plants support each other's needs (e.g., the Three Sisters: corn, beans, and squash).
- Incorporate animals into your garden to assist with tasks such as pest control, soil fertility, and waste recycling.

9. Use Small and Slow Solutions

Principle Overview:
- Small and slow solutions are often more sustainable and manageable than large, rapid changes. They allow for careful observation, adaptation, and improvement over time.
- This principle emphasizes incremental development and long-term thinking.

Application:
- Start with small garden projects and expand gradually as you gain experience and confidence.
- Use slow-growing, perennial plants that establish stable ecosystems over time.

- Implement water-saving techniques like drip irrigation and mulching, which are gradual but highly effective.

10. Use and Value Diversity

Principle Overview:
- Diversity in plant species, animals, and practices increases resilience and reduces vulnerability to pests, diseases, and environmental changes.
- This principle promotes polyculture over monoculture, encouraging a variety of elements that complement and support each other.

Application:
- Plant a diverse range of crops and varieties to ensure continuous harvests and reduce the risk of total crop failure.
- Create habitat diversity by incorporating different plant heights, structures, and functions in your garden.

- Use crop rotation and intercropping to maintain soil health and prevent pest and disease buildup.

11. Use Edges and Value the Marginal

Principle Overview:
- Edges – the boundaries between different ecosystems – are often the most productive and diverse areas. Valuing and utilizing these edges maximizes productivity.
- Marginal areas, which may be overlooked or underutilized, can offer unique opportunities for growth and innovation.

Application:
- Design garden beds and ponds with irregular shapes to increase edge effects and biodiversity.
- Utilize the edges of pathways, fences, and structures for planting productive crops and flowers.

- Experiment with marginal areas, such as shaded corners or steep slopes, to find suitable plants and uses for these spaces.

12. Creatively Use and Respond to Change

Principle Overview:
- Change is inevitable in natural systems. Embracing and adapting to change allows you to turn challenges into opportunities and maintain a resilient garden.
- This principle encourages flexibility, innovation, and a proactive approach to problem-solving.

Application:
- Monitor environmental changes and adjust your garden practices accordingly. For example, adapt planting schedules based on shifting weather patterns.
- Use succession planting to respond to seasonal changes and ensure continuous production.

- Experiment with new techniques and crops to diversify your garden and stay ahead of potential challenges.

Conclusion

Permaculture design principles provide a holistic framework for creating sustainable, resilient, and productive gardens. By understanding and applying these principles, you can design systems that work in harmony with nature, maximize yields, and minimize waste. Each principle offers valuable insights and strategies that, when integrated into your permaculture practice, can transform your garden into a thriving ecosystem. Embrace these principles as guiding lights on your permaculture journey, and continually observe, learn, and adapt to create a garden that sustains both you and the environment.

Zoning and Sector Analysis

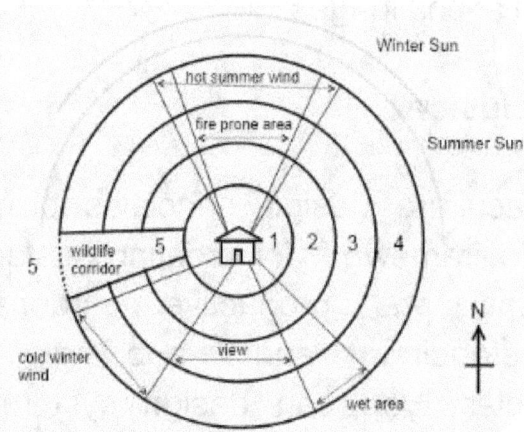

Zoning and Sector Analysis: A Comprehensive Guide for Permaculture Gardening

Understanding zoning and sector analysis is crucial in permaculture design to optimize the use of space, energy, and resources in your garden. These tools help you strategically place elements in your garden to maximize efficiency, productivity, and sustainability. This guide explores zoning and sector analysis in detail, offering insights into their application and importance in permaculture gardening.

Understanding Zoning in Permaculture

Definition of Zoning:
- Zoning is a method of organizing your garden into distinct areas based on how frequently different elements are used and maintained.
- Zones range from Zone 0 (the home) to Zone 5 (wilderness), with each zone serving specific functions and hosting particular activities.

The Five Zones:

1. Zone 0: The Home:
   - This is the living space and the heart of your permaculture system.
   - Activities: Cooking, eating, living, and initial processing of food.
   - Elements: Kitchen garden, herbs, indoor plants, and composting systems.

2. Zone 1: The Immediate Surroundings:

- This zone includes areas closest to the home, visited multiple times daily.
- Activities: Daily harvesting, quick-access vegetable beds, and frequent maintenance.
- Elements: Salad greens, culinary herbs, compost bins, rainwater collection systems, and small livestock (e.g., chickens).

3. Zone 2: The Garden:

- Located slightly further from the home, this zone requires regular attention but not daily visits.
- **Activities:** Larger scale vegetable gardening, weekly maintenance, and perennial crops.
- Elements: Fruit trees, larger vegetable plots, beehives, and small ponds.

4. Zone 3: The Farm:

- This zone is for less frequently maintained areas, visited weekly or monthly.
- **Activities:** Staple crop production, orchards, and larger livestock management.

- Elements: Grain fields, orchards, pastures, and larger water bodies.

5. Zone 4: The Semi-Wild:
- A managed wild area that requires minimal intervention, visited seasonally.
- Activities: Foraging, fuelwood harvesting, and wildlife habitat management.
- Elements: Woodlots, wild food crops, and unmanaged pasture.

6. Zone 5: The Wilderness:
- This zone is left completely wild, providing a natural reserve and a source of inspiration.
- **Activities**: Observation, research, and conservation.
- **Elements:** Native forest, wetlands, and natural ecosystems.

Applying Zoning Principles:
- Assess the needs and habits of your household to determine the appropriate placement of elements.

- Design your garden so that frequently used and maintained elements are placed in closer zones.
- Utilize pathways and access routes efficiently to reduce time and energy spent on maintenance.

Understanding Sector Analysis in Permaculture

Definition of Sector Analysis:
- Sector analysis examines the external forces and energies that affect your garden, such as sunlight, wind, water flow, and wildlife.
- It helps you design your garden to harness beneficial energies and mitigate negative impacts.

Key Sectors to Analyze:

1. Sunlight:
 - Identify the direction and intensity of sunlight throughout the day and across seasons.
 - Use this information to place sun-loving plants, shade-tolerant plants, and structures accordingly.

2. Wind:
 - Determine the prevailing wind directions and intensity.
 - Design windbreaks, plant hedges, or place barriers to protect delicate plants and structures from strong winds.

3. Water Flow:
 - Analyze the natural flow of water through your landscape, including rainwater runoff and groundwater movement.
 - Implement water harvesting techniques like swales, ponds, and rain gardens to capture and utilize water efficiently.

4. Wildlife:
- Observe patterns of wildlife movement and behavior in and around your garden.
- Create habitats and corridors for beneficial wildlife while implementing strategies to deter pests.

5. Frost and Fire:
- Identify areas prone to frost pockets or fire risk.
- Use frost-resistant plants in vulnerable areas and design firebreaks to protect your garden from wildfire.

Applying Sector Analysis Principles:
- Create a sector map of your garden, indicating the direction and influence of each sector.
- Design your garden layout to take advantage of positive forces (e.g., sunlight, beneficial winds) and mitigate negative forces (e.g., strong winds, water runoff).

- Place elements in locations that optimize their performance and resilience based on sector influences.

Integrating Zoning and Sector Analysis

Holistic Design Approach:
- Combine zoning and sector analysis to create a comprehensive design plan that maximizes efficiency and sustainability.
- Use zoning to organize elements based on frequency of use and maintenance needs, and sector analysis to optimize their placement based on external influences.

Practical Steps for Integration:
1. Mapping and Observation:
    - Start by creating a base map of your garden, including existing elements and features.
    - Conduct thorough observations over time to gather data on sunlight, wind, water flow, wildlife, frost, and fire sectors.

2. Zoning:
 - Divide your garden into zones based on the frequency of use and maintenance. Consider the proximity to the home and the types of activities performed in each area.
 - Assign specific elements to appropriate zones, ensuring that frequently accessed elements are placed closer to the home.

3. Sector Analysis:
 - Overlay your base map with sector information, indicating the direction and intensity of external forces.
 - Analyze how these forces interact with each zone and the elements within them.

4. Design Implementation:
 - Adjust the placement of elements based on both zoning and sector analysis. For example, place a vegetable garden (Zone 1) in a sunny spot with wind protection, or locate water catchment systems (Zone 2 or 3) where water naturally flows.

- Use natural features and constructed elements to enhance positive sector influences and mitigate negative ones. For example, plant trees to create shade and windbreaks or build swales to manage water flow.

5. Iterative Process:

- Design is an ongoing process. Continuously observe and interact with your garden, adjusting the placement of elements and refining your design based on feedback and changing conditions.
- Use permaculture principles such as "Observe and Interact" and "Creatively Use and Respond to Change" to adapt and improve your garden over time.

Conclusion

Zoning and sector analysis are fundamental tools in permaculture design, allowing you to create a garden that is efficient, productive, and sustainable. By organizing your garden into zones based on usage and maintenance needs, and by analyzing external forces through sector analysis, you can optimize the placement of elements to enhance their performance and resilience. This integrated approach ensures that your garden works harmoniously with natural systems, conserving energy and resources while maximizing yields. Embrace the principles of zoning and sector analysis to transform your garden into a thriving permaculture system that supports both human and ecological health.

Designing for Sustainability

Designing for Sustainability in Permaculture Gardening

Sustainability is at the heart of permaculture, aiming to create systems that are ecologically balanced, economically viable, and socially just. Designing for sustainability involves thoughtful planning and implementation of practices that ensure long-term productivity and resilience while minimizing negative environmental impacts. This comprehensive guide explores key strategies and principles for designing a sustainable permaculture garden.

Core Principles of Sustainable Design

1. Systems Thinking:
- Approach your garden as an interconnected system where every element supports and enhances the others.
- Understand the relationships between plants, animals, soil, water, and human activity to design integrated systems that function cohesively.

2. Resource Efficiency:
- Maximize the use of renewable resources and minimize waste.

- Implement techniques to conserve water, energy, and materials, ensuring resources are used efficiently and sustainably.

3. Biodiversity:
- Foster a diverse range of plant and animal species to create resilient ecosystems.
- Diversity enhances ecological stability, reduces the risk of pest and disease outbreaks, and supports pollinators and beneficial insects.

4. Resilience and Adaptability:
- Design systems that can withstand and recover from environmental stresses and changes.
- Incorporate flexible and adaptable practices to ensure long-term sustainability in the face of climate change and other challenges.

5. Ethics and Values:
- Adhere to the core permaculture ethics of Earth Care, People Care, and Fair Share.

- Ensure that your design practices are equitable, inclusive, and respect natural ecosystems.

Key Strategies for Designing a Sustainable Permaculture Garden

1. Soil Health and Fertility

Building and Maintaining Healthy Soil:
- **Composting:** Create rich, organic compost from kitchen scraps, garden waste, and animal manure. Compost improves soil structure, fertility, and microbial activity.
- **Mulching:** Apply organic mulch to retain soil moisture, suppress weeds, and add nutrients as it decomposes. Mulch also protects soil from erosion and temperature fluctuations.
- **Cover Cropping**: Grow cover crops like clover, vetch, or rye to improve soil fertility, structure, and organic matter content. Cover crops prevent soil erosion and suppress weeds.

- **No-Till Gardening**: Minimize soil disturbance to preserve soil structure, microbial communities, and organic matter. No-till practices reduce erosion and improve soil health over time.

Nutrient Cycling:
- Chop and Drop: Use the "chop and drop" technique to prune plants and leave the cuttings on the soil surface. This method adds organic matter and nutrients back into the soil.
- **Green Manure:** Grow and incorporate green manure crops to add nitrogen and organic matter to the soil. Green manures enhance soil fertility and structure.

2. Water Management

Efficient Water Use:
- Rainwater Harvesting: Collect and store rainwater using barrels, tanks, or cisterns. Use harvested rainwater for irrigation and other garden needs.

- **Drip Irrigation:** Implement drip irrigation systems to deliver water directly to plant roots, minimizing evaporation and runoff. Drip irrigation is highly efficient and conserves water.
- **Greywater Systems:** Recycle household greywater (from sinks, showers, and washing machines) for garden irrigation. Ensure greywater is filtered and used safely.

Water Conservation Techniques:
- Mulching: Mulch helps retain soil moisture, reducing the need for frequent watering.
- **Swales and Contour Planting:** Design swales and plant along contours to capture and slow down rainwater, allowing it to infiltrate the soil. These techniques reduce runoff and erosion.
- **Rain Gardens:** Create rain gardens to capture and filter rainwater, reducing runoff and promoting groundwater recharge.

3. Plant Selection and Biodiversity

Choosing the Right Plants:
- **Native and Adapted Species:** Select native plants and those adapted to your local climate and soil conditions. Native plants are well-suited to local conditions and support local wildlife.
- **Perennials:** Plant perennials to provide long-term yields and reduce the need for replanting. Perennials contribute to soil stability and biodiversity.

Promoting Biodiversity:
- **Polyculture**: Grow a mix of plant species in the same area to mimic natural ecosystems. Polyculture enhances resilience, reduces pest and disease pressure, and increases yields.
- **Plant Guilds:** Design plant guilds, where groups of plants support each other's growth. For example, a typical guild includes a nitrogen-fixing plant, a dynamic accumulator, and a ground cover.

- **Habitat Creation**: Create diverse habitats such as ponds, hedgerows, and wildflower meadows to support beneficial insects, birds, and other wildlife.

4. Renewable Energy and Sustainable Infrastructure

Energy Efficiency:
- Passive Solar Design: Use passive solar principles to design garden structures that maximize natural light and heat. For example, orient greenhouses and cold frames to capture sunlight.
- **Solar Panels and Wind Turbines:** Install renewable energy systems to power garden tools, lighting, and irrigation pumps. Renewable energy reduces reliance on fossil fuels.

Sustainable Building Materials:
- **Natural and Recycled Materials**: Use natural materials like wood, stone, and bamboo, or recycled materials for garden

structures. Sustainable materials reduce environmental impact and support local economies.

- **Low-Impact Construction**: Implement construction practices that minimize soil disturbance, pollution, and resource consumption. Use eco-friendly building techniques and materials.

5. Waste Reduction and Recycling

Zero Waste Practices:
- Composting: Recycle organic waste into compost to enrich your soil. Composting reduces landfill waste and greenhouse gas emissions.
- **Vermiculture:** Use worm composting (vermiculture) to convert kitchen waste into nutrient-rich worm castings. Worm composting is efficient and produces high-quality soil amendments.
- **Reusing Materials:** Repurpose old containers, wood pallets, and other

materials for garden use. Reusing materials reduces waste and saves money.

Resourceful Gardening:
- **Seed Saving:** Save seeds from your plants to ensure a continuous supply of locally adapted varieties. Seed saving promotes genetic diversity and resilience.
- **Upcycling:** Use creativity to upcycle materials into functional garden items, such as turning old tires into planters or using glass bottles for garden edging.

6. Community Engagement and Education

Building Community Resilience:
- Community Gardens: Establish community gardens to promote local food production, social interaction, and education. Community gardens foster a sense of ownership and cooperation.
- Skill Sharing: Host workshops, classes, and events to share knowledge and skills

about permaculture and sustainable living. Education empowers individuals and strengthens communities.

Social Sustainability:
- **Fair Share Principle**: Ensure that resources and benefits are distributed fairly within your community. Practice generosity and cooperation to build a supportive network.
- **Inclusivity and Accessibility:** Design your garden to be accessible and inclusive, accommodating people of all abilities and backgrounds. Inclusivity fosters community engagement and well-being.

Conclusion

Designing for sustainability in permaculture gardening involves integrating ecological principles, resource efficiency, biodiversity, and community engagement. By focusing on soil health, water management, plant selection, renewable energy, waste reduction, and social sustainability, you can create a resilient and productive garden that supports both human and ecological health. Embrace these strategies and principles to design a permaculture garden that thrives sustainably, providing long-term benefits for you and the environment. Through thoughtful planning and ongoing adaptation, your permaculture garden can become a model of sustainability, contributing to a healthier and more resilient world.

CHAPTER 2:

Soil Health

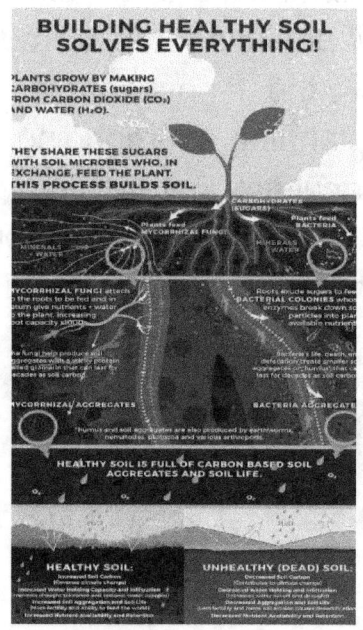

Comprehensive Guide to Soil Health

Soil health, also known as soil quality, is the capacity of soil to function as a living ecosystem that sustains plants, animals, and humans. Understanding and maintaining soil health is critical for agriculture, environmental sustainability, and food security. This guide explores the components, significance, and practices for enhancing soil health.

Components of Soil Health

1. Physical Properties:

- Soil Texture: The relative proportion of sand, silt, and clay particles. Texture affects water retention, drainage, and root penetration.

- **Soil Structure:** The arrangement of soil particles into aggregates. Good structure enhances porosity, water infiltration, root growth, and resistance to erosion.

- **Bulk Density:** The mass of soil per unit volume. Lower bulk density indicates better soil aeration and root penetration.

- **Soil Porosity:** The volume of pore spaces in soil. High porosity improves water infiltration, drainage, and root respiration.

2. Chemical Properties:

- **Soil pH:** Measures the acidity or alkalinity of soil. Most plants prefer a pH of 6.0 to 7.5, where nutrient availability is optimal.
- **Nutrient Content:** Essential nutrients include macronutrients (nitrogen, phosphorus, potassium) and micronutrients (iron, manganese, zinc).
- **Cation Exchange Capacity (CEC):** The soil's ability to hold and exchange cations. High CEC soils can retain and supply more nutrients to plants.
- **Salinity:** The concentration of soluble salts in soil, which can affect plant growth negatively at high levels.

3. Biological Properties:

- **Soil Microorganisms:** Bacteria, fungi, protozoa, and nematodes play crucial roles in decomposing organic matter, nutrient cycling, and disease suppression.

- **Soil Fauna:** Earthworms and insects contribute to soil aeration and organic matter decomposition.

- **Soil Organic Matter (SOM):** Decomposed plant and animal residues. SOM improves soil structure, nutrient availability, and water retention.

Importance of Soil Health

1. Plant Growth and Productivity:
- Healthy soil provides essential nutrients, water, and physical support for plants, leading to vigorous growth and high yields.
- Poor soil health can result in nutrient deficiencies, stunted growth, and increased susceptibility to pests and diseases.

2. Water Management:
- Healthy soil improves water infiltration and retention, reducing runoff and erosion.
- Well-structured soil helps in maintaining adequate moisture levels for plants, especially during drought.

3. Carbon Sequestration:
   - Soils act as significant carbon sinks. Healthy soils with high organic matter content can sequester more carbon, mitigating climate change.

4. Biodiversity and Ecosystem Services:
   - Soil biodiversity supports nutrient cycling, pest and disease suppression, and organic matter decomposition.
   - Diverse soil organisms contribute to overall ecosystem resilience and productivity.

5. Environmental Protection:
   - Healthy soils filter pollutants, reducing the risk of water contamination.
   - They also mitigate the impacts of extreme weather events by enhancing water infiltration and reducing erosion.

Strategies for Maintaining and Improving Soil Health

1. Organic Matter Management:
   - Composting: Add compost to increase organic matter, improve soil structure, and enhance nutrient availability.
   - **Mulching:** Apply organic mulches to conserve soil moisture, suppress weeds, and add organic matter as they decompose.
   - **Green Manures and Cover Crops**: Plant cover crops like clover or rye to add organic matter, fix nitrogen, and protect soil from erosion.

2. Soil Structure and Tillage:
   - Reduced Tillage: Minimize soil disturbance to preserve soil structure and microbial activity.
   - **No-Till Farming:** Avoid tilling to maintain soil structure, reduce erosion, and improve water retention.

- **Contour Farming:** Plant along natural contours to reduce water runoff and soil erosion.

3. Nutrient Management:
- Soil Testing: Regularly test soil to monitor nutrient levels and pH. Use the results to guide fertilization practices.
- **Balanced Fertilization**: Apply fertilizers based on soil test recommendations to prevent nutrient imbalances and environmental harm.
- **Organic Amendments**: Use compost, manure, and other organic materials to improve nutrient availability and soil health.

4. Enhancing Soil Microbial Activity:
- **Compost Tea**: Apply compost tea to introduce beneficial microorganisms into the soil.
- **Mycorrhizal Fungi**: Inoculate soil with mycorrhizal fungi to enhance nutrient uptake and improve soil structure.

- **Avoiding Synthetic Chemicals:** Minimize or eliminate the use of synthetic pesticides and fertilizers that can harm beneficial soil organisms.

5. Erosion Control:
- **Cover Crops**: Use cover crops to protect soil from erosion and improve soil structure.
- **Mulching:** Apply mulches to prevent soil erosion and enhance soil moisture retention.
- **Terracing:** Build terraces on sloped land to reduce runoff and soil erosion.

6. Water Management:
- **Efficient Irrigation**: Use drip irrigation to minimize water wastage and ensure efficient water delivery to plant roots.
- **Rainwater Harvesting**: Collect and store rainwater for irrigation purposes.
- **Greywater Recycling**: Recycle household greywater for garden use, ensuring it is treated appropriately.

7. Integrated Pest Management (IPM):

- **Biological Controls**: Use natural predators and beneficial insects to control pests.

- **Companion Planting**: Grow plants that repel pests or attract beneficial insects.

- **Cultural Practices**: Rotate crops and use diverse planting strategies to reduce pest pressure.

8. Climate Considerations:

- **Climate-Resilient Plants**: Choose plant varieties suited to local climate conditions.

- **Microclimates**: Create microclimates using windbreaks and shade structures to protect plants from extreme weather.

Monitoring and Assessing Soil Health

1. Regular Soil Testing:

- Conduct soil tests every 2-3 years to monitor nutrient levels, pH, and organic matter content. Adjust management practices based on the results.

2. Visual and Physical Assessment:
- Observe soil color, texture, structure, and root development. Healthy soil is dark, crumbly, and well-aggregated with visible roots and earthworms.

3. Biological Indicators:
- Monitor the presence and activity of soil organisms. High biodiversity and abundant soil life indicate healthy soil.

Practical Examples and Techniques

1. Composting:
- Setup: Create a compost bin or pile using green materials (kitchen scraps, grass clippings) and brown materials (leaves, straw).
- **Maintenance**: Turn the compost regularly to aerate it and speed up decomposition. Finished compost should be dark, crumbly, and earthy-smelling.
- **Application:** Apply compost to garden beds to improve soil fertility and structure.

2. Mulching:

- **Materials**: Use organic materials such as straw, leaves, wood chips, or grass clippings.
- **Application**: Apply a thick layer (3-4 inches) around plants and over garden beds to conserve moisture, suppress weeds, and add organic matter.
- **Decomposition**: Mulch gradually decomposes, enriching the soil with organic matter.

3. Cover Cropping:

- **Planting:** Grow cover crops during the off-season or in fallow areas. Common cover crops include clover, vetch, and rye.
- **Management:** Mow or cut cover crops before they set seed, and incorporate them into the soil as green manure to add organic matter and improve soil structure.
- **Benefits:** Cover crops protect soil from erosion, enhance soil fertility, and improve soil structure.

4. No-Till Gardening:

- **Method:** Use a broadfork to aerate the soil without turning it over, preserving soil structure and microbial communities.
- **Advantages**: Reduces erosion, improves water retention, and enhances soil organic matter.

5. Crop Rotation:

- **Planning**: Rotate crops to prevent nutrient depletion, reduce pest and disease buildup, and improve soil structure. Plan rotations based on plant families and nutrient requirements.

Conclusion

Soil health is fundamental to sustainable agriculture and environmental protection. By understanding and managing the physical, chemical, and biological properties of soil, we can enhance its ability to support plant growth, regulate water, filter pollutants, and sustain biodiversity. Implementing practices such as adding organic matter, minimizing soil disturbance, managing nutrients, and promoting biodiversity can significantly improve soil health. Regular monitoring and assessment ensure that soil management practices are effective and sustainable, contributing to long-term agricultural productivity and ecological resilience.

1. *Building Healthy Soil*

Building Healthy Soil: A Comprehensive Guide

Creating and maintaining healthy soil is fundamental for sustainable gardening and agriculture. Healthy soil supports plant growth, maintains ecological balance, and ensures long-term productivity. This comprehensive guide covers the principles, practices, and benefits of building healthy soil.

Principles of Building Healthy Soil

1. Enhancing Soil Structure:

- Aim for a loose, crumbly soil texture that allows for proper root growth and water infiltration.
- Encourage the formation of soil aggregates, which improve soil porosity and resilience against erosion.

2. Increasing Organic Matter:

- Organic matter improves soil structure, water retention, and nutrient availability.
- It serves as a food source for soil microorganisms, promoting a healthy soil ecosystem.

3. Balancing Soil Nutrients:

- Ensure a balanced supply of essential nutrients for plants.
- Avoid nutrient imbalances that can harm plant health and soil microorganisms.

4. Promoting Biodiversity:
- Encourage a diverse community of soil organisms, including bacteria, fungi, earthworms, and insects.
- Biodiversity enhances nutrient cycling, disease suppression, and soil structure.

5. Minimizing Soil Disturbance:
- Reduce tillage and other practices that disrupt soil structure and microbial communities.
- Preserve soil integrity and promote natural processes.

6. Managing Water Wisely:
- Maintain optimal soil moisture levels to support plant growth and microbial activity.
- Prevent waterlogging and drought stress.

Practices for Building Healthy Soil

1. Adding Organic Matter:

   - **Composting:** Create and use compost from kitchen scraps, garden waste, and other organic materials.

   - **Setup:** Combine green materials (e.g., vegetable scraps) and brown materials (e.g., leaves) in a compost bin.

      - **Maintenance:** Turn the compost regularly to aerate and speed up decomposition.

   - **Application:** Apply finished compost to garden beds to enrich the soil.

   - **Mulching:** Apply organic mulch (e.g., straw, wood chips) to garden beds.

      - **Benefits**: Mulch conserves moisture, suppresses weeds, and adds organic matter as it decomposes.

   - **Green Manures and Cover Crops:** Plant cover crops such as clover, rye, or vetch.

- **Planting**: Grow cover crops during off-seasons or in fallow areas.

- **Management**: Cut and incorporate cover crops into the soil to add organic matter and improve soil structure.

2. Improving Soil Structure:

- **No-Till Gardening**: Avoid tilling the soil to preserve its structure and microbial communities.

- **Method**: Use a broadfork or similar tool to aerate the soil without turning it over.

- **Contour Farming**: Plant along natural land contours to reduce soil erosion and water runoff.

- **Raised Beds:** Use raised beds to improve drainage and soil structure, especially in areas with poor soil.

3. Balancing Soil Nutrients:

- **Soil Testing**: Regularly test soil to determine nutrient levels and pH.

- **Adjustment**: Use soil test results to guide fertilization and amendment practices.

- **Organic Fertilizers**: Apply organic fertilizers (e.g., compost, manure) to provide essential nutrients.
- **Benefits:** Organic fertilizers improve soil health and reduce the risk of nutrient leaching.

- **Crop Rotation:** Rotate crops to prevent nutrient depletion and reduce pest and disease buildup.
- **Planning:** Alternate crops with different nutrient requirements and root structures.

4. Promoting Soil Biodiversity:
- **Beneficial Microorganisms**: Inoculate soil with beneficial microorganisms, such as mycorrhizal fungi and compost tea.
- **Mycorrhizal Fungi**: Enhance nutrient uptake and improve soil structure.

- **Compost Tea**: Brew and apply compost tea to introduce beneficial microorganisms.

- **Plant Diversity**: Grow a variety of plants to support diverse soil organisms.

- **Companion Planting:** Plant complementary species that benefit each other.

- **Reduced Chemical Use:** Minimize the use of synthetic pesticides and fertilizers that can harm beneficial soil organisms.

5. Water Management:
- **Efficient Irrigation**: Use drip irrigation or soaker hoses to provide water directly to plant roots.

- **Benefits**: Reduces water wastage and prevents waterlogging.

- **Rainwater Harvesting**: Collect and store rainwater for irrigation.

- **Setup**: Use barrels or tanks to capture and store rainwater from roofs and other surfaces.

- **Mulching**: Apply mulch to conserve soil moisture and reduce evaporation.

6. Erosion Control:
   - **Cover Crops**: Plants cover crops to protect soil from erosion.
      - **Benefits:** Cover crops improve soil structure and add organic matter.

- **Mulching**: Use mulch to protect soil surfaces from wind and water erosion.
- **Terracing**: Build terraces on sloped land to reduce runoff and soil erosion.

Benefits of Healthy Soil

1. Enhanced Plant Growth and Yield:
   - Healthy soil provides optimal conditions for root growth, nutrient uptake, and water availability.

- Plants grown in healthy soil are more vigorous, productive, and resilient.

2. Improved Water Retention and Drainage:
- Good soil structure and organic matter content enhance water infiltration and retention.
- Healthy soil prevents waterlogging and drought stress.

3. Increased Nutrient Availability:
- Healthy soil has a balanced supply of essential nutrients and a high cation exchange capacity (CEC).
- Organic matter and microbial activity improve nutrient cycling and availability.

4. Resilience to Pests and Diseases:
- Biodiverse soil ecosystems suppress soil-borne pests and diseases.
- Healthy plants are more resistant to pests and diseases.

5. Carbon Sequestration:
- Soils with high organic matter content sequester more carbon, mitigating climate change.
- Healthy soil practices contribute to long-term carbon storage.

6. Ecosystem Services:
- Healthy soil supports biodiversity, nutrient cycling, and water filtration.
- It contributes to overall ecosystem resilience and stability.

Monitoring Soil Health

1. Regular Soil Testing:
- Test soil every 2-3 years to monitor nutrient levels, pH, and organic matter content.

- Use the results to adjust soil management practices.

2. Visual and Physical Assessment:
- Observe soil color, texture, structure, and root development.
- Healthy soil is dark, crumbly, and well-aggregated with visible roots and earthworms.

3. Biological Indicators:
- Monitor the presence and activity of soil organisms.
- High biodiversity and abundant soil life indicate healthy soil.

Challenges and Solutions in Building Healthy Soil

1. Compaction:
- **Problem**: Soil compaction reduces pore space, limiting root growth and water infiltration.

- **Solution:** Avoid heavy machinery and excessive foot traffic on wet soil. Use cover crops and organic matter to improve soil structure.

2. Erosion:

- **Problem:** Soil erosion removes topsoil, organic matter, and nutrients.
- **Solution:** Use cover crops, mulch, and terracing to protect soil from erosion.

3. Nutrient Imbalances:

- **Problem:** Excessive or deficient nutrients can harm plants and soil organisms.
- **Solution:** Conduct soil tests and apply fertilizers based on recommendations. Use organic amendments to improve nutrient availability.

4. Salinity:

- **Problem:** High soil salinity can damage plant roots and reduce growth.

- **Solution**: Improve drainage, use salt-tolerant plants, and apply gypsum to reduce soil salinity.

5. Pest and Disease Pressure:
- **Problem**: Soil-borne pests and diseases can reduce plant health and yields.
- **Solution**: Promote soil biodiversity, rotate crops, and use organic pest control methods.

Conclusion

Building healthy soil is a continuous process that requires understanding its physical, chemical, and biological properties. By adding organic matter, improving soil structure, balancing nutrients, promoting biodiversity, managing water wisely, and controlling erosion, gardeners and farmers can create and maintain productive, sustainable soils. Healthy soil supports robust plant growth, enhances ecosystem services, and contributes to environmental sustainability. Regular monitoring and adaptive management ensure that soil health practices are effective and sustainable, leading to long-term agricultural productivity and ecological resilience.

Composting Basics

Composting Basics: A Comprehensive Guide

Composting is the natural process of recycling organic material, such as leaves and food scraps, into a rich soil amendment known as compost. Composting reduces waste, enriches the soil, and promotes sustainable gardening and farming. This guide covers the essentials of composting, including the science behind it, materials, methods, benefits, and troubleshooting common issues.

The Science of Composting

Composting is an aerobic (oxygen-requiring) process that involves the decomposition of organic matter by microorganisms, such as bacteria and fungi. These microorganisms break down organic material into humus, a stable, nutrient-rich substance.

Stages of Composting:
1. Mesophilic Phase (15-40°C): Initial stage where mesophilic microorganisms,

which thrive at moderate temperatures, begin breaking down organic matter.

2. Thermophilic Phase (40-70°C): As the compost heats up, thermophilic microorganisms take over. This phase rapidly decomposes organic materials and kills pathogens and weed seeds.

3. Cooling Phase: The compost pile gradually cools down, and mesophilic microorganisms resume activity to further decompose the material.

4. Maturation Phase: The compost stabilizes and matures, resulting in a dark, crumbly, and earthy-smelling product.

Materials for Composting

Composting requires a balance of carbon-rich and nitrogen-rich materials, often referred to as "browns" and "greens," respectively.

Carbon-Rich Materials (Browns):
- Dry leaves

- Straw or hay
- Wood chips and sawdust (untreated)
- Shredded newspaper and cardboard
- Corn stalks

Nitrogen-Rich Materials (Greens):
- Vegetable and fruit scraps
- Coffee grounds and tea bags
- Grass clippings
- Manure from herbivores (e.g., cows, horses, rabbits)
- Green leaves and plant trimmings

Avoid These Materials:
- Meat, dairy, and oily foods (attract pests and cause odors)
- Diseased plants (may spread diseases)
- Weeds with seeds (may germinate in compost)
- Pet waste from carnivores (contains harmful pathogens)
- Treated wood or sawdust (contains chemicals)

Methods of Composting

1. Traditional Compost Pile:

- **Setup**: Create a compost pile directly on the ground, preferably in a shady spot.
- **Layering**: Alternate layers of browns and greens, aiming for a ratio of about 3:1 (browns to greens).
- **Turning:** Turn the pile regularly to aerate it and speed up decomposition.
- **Moisture**: Keep the pile moist, but not waterlogged. It should feel like a damp sponge.

2. Compost Bins:

- Types: Choose from stationary bins, tumblers, or homemade bins.
- **Advantages**: Bins contain the compost, making it neater and easier to manage. Tumblers simplify turning the compost.
- **Usage:** Add materials as you generate them, and turn or mix the compost regularly.

3. Vermicomposting:

- **Worms**: Use red worms (Eisenia fetida) to decompose organic matter.

- Setup: Create a worm bin with bedding material (e.g., shredded paper, coconut coir) and add food scraps.

- **Care:** Maintain moisture, avoid overfeeding, and harvest worm castings regularly.

4. Trench or Pit Composting:

- Method: Dig a trench or pit, add organic materials, and cover with soil.

- **Advantages**: Minimal effort and good for adding nutrients directly to garden beds.

- **Considerations**: Decomposition takes longer and may attract pests if not buried deeply.

The Composting Process

1. Site Selection and Setup:
   - Choose a level, well-drained site with some shade to prevent the compost from drying out.
   - Ensure the area is accessible for adding materials and turning the compost.

2. Building the Pile:
   - Start with a layer of coarse materials, such as straw or small branches, to aid aeration.
   - Alternate layers of browns and greens, aiming for a mix of carbon and nitrogen-rich materials.
   - Moisten each layer as you build the pile to maintain moisture levels.

3. Maintaining the Compost:
   - **Turning**: Turn the pile every 1-2 weeks to introduce oxygen and speed up decomposition.

- **Moisture**: Keep the compost moist but not soggy. Add water during dry periods and cover the pile during heavy rains.
- **Temperature:** Monitor the temperature. Active composting should generate heat (up to 60-70°C) in the thermophilic phase.

4. Maturation:
- Allow the compost to mature for several weeks to months after the active phase.
- The compost is ready when it is dark, crumbly, and has an earthy smell.

Benefits of Composting

1. Soil Improvement:
- Compost adds organic matter, improving soil structure, water retention, and aeration.
- It enhances nutrient availability and supports healthy plant growth.

2. Nutrient Recycling:

- Compost returns valuable nutrients to the soil, reducing the need for chemical fertilizers.
- It contains essential nutrients, including nitrogen, phosphorus, and potassium, in a slow-release form.

3. Waste Reduction:

- Composting diverts organic waste from landfills, reducing methane emissions and landfill use.
- It promotes sustainable waste management practices.

4. Environmental Benefits:

- Composting reduces the need for chemical fertilizers and pesticides, protecting water quality and biodiversity.
- It sequesters carbon in the soil, mitigating climate change.

5. Economic Benefits:
- Using compost can reduce gardening and landscaping costs by lowering the need for fertilizers, soil amendments, and water.

Troubleshooting Common Issues

1. Odors:
- Problem: Unpleasant odors indicate anaerobic conditions or too much nitrogen.
- Solution: Turn the compost to aerate it and add more carbon-rich materials.

2. Slow Decomposition:
- Problem: The compost pile is not heating up or decomposing slowly.
- **Solution:** Check the carbon-to-nitrogen ratio and moisture levels. Ensure the pile is large enough to retain heat.

3. Pests:
- Problem: Rodents, flies, or other pests are attracted to the compost.

- **Solution:** Avoid adding meat, dairy, and oily foods. Cover food scraps with browns and turn the pile regularly.

4. Excess Moisture:
   - Problem: The compost is too wet and soggy.
   - Solution: Add dry carbon-rich materials, turn the pile to aerate it, and cover it during heavy rains.

5. Dry Compost:
   - Problem: The compost is too dry and not decomposing.
   - Solution: Add water to moisten the pile and turn it to distribute moisture evenly.

Advanced Composting Techniques

1. Hot Composting:
   - Method: Create a large pile (at least 1 cubic meter) to generate and retain heat.
   - **Benefits:** Rapid decomposition and pathogen/weed seed elimination.

- **Maintenance:** Turn the pile frequently to maintain high temperatures.

2. Cold Composting:
- Method: Add materials gradually without turning the pile frequently.
- **Benefits:** Less labor-intensive, but decomposition takes longer.
- **Considerations**: Ensure a good mix of browns and greens to prevent odors.

3. Bokashi Composting:
- Method: Use anaerobic fermentation with a special inoculant (bokashi bran) to decompose food waste.
- **Setup**: Place food scraps in an airtight container, add bokashi bran, and let it ferment.
- **Usage:** Bury the fermented material in soil to complete decomposition.

4. Sheet Composting:

- **Method:** Spread organic materials directly on garden beds and cover with mulch or soil.
- **Benefits:** Builds soil organic matter and suppresses weeds.
- **Considerations:** Allow time for materials to decompose before planting.

Conclusion

Composting is an essential practice for sustainable gardening and farming. By understanding the science of composting, selecting the right materials, and choosing appropriate methods, anyone can create high-quality compost to improve soil health and reduce waste. Regular maintenance, monitoring, and troubleshooting ensure successful composting and the production of a valuable soil amendment. With benefits ranging from enhanced plant growth to environmental protection, composting is a key component of sustainable land management practices.

Mulching and Cover Crops

Mulching and Cover Crops: A Comprehensive Guide to Sustainable Soil Management

Mulching and cover crops are fundamental techniques in sustainable gardening and agriculture, contributing significantly to soil health, moisture conservation, weed control, and overall ecosystem balance. These practices mimic natural processes, enhancing the resilience and productivity of gardens and farms. This comprehensive guide explores the principles, types, benefits, and practical applications of mulching and cover crops.

Mulching: Principles and Practices

Mulching involves covering the soil with organic or inorganic materials to improve soil health and plant growth. This practice offers multiple benefits, including moisture retention, temperature regulation, weed suppression, and enhanced soil fertility.

Types of Mulch

1. Organic Mulch:
   - Compost: Enriches the soil with nutrients and improves its structure.

- Straw or Hay: Ideal for vegetable gardens; decomposes slowly and adds organic matter.

- **Wood Chips and Bark:** Suitable for perennial beds and pathways; decomposes slowly, providing long-term benefits.

- **Grass Clippings:** High in nitrogen; excellent for garden beds but should be applied in thin layers to prevent matting.

- **Leaf Mold:** Decomposed leaves that improve soil structure and moisture retention.

- **Pine Needles:** Acidic mulch suitable for acid-loving plants like blueberries and azaleas.

2. Inorganic Mulch:

- Plastic Mulch: Used in commercial farming for weed control and soil warming.

- **Landscape Fabric:** A breathable material that suppresses weeds while allowing water and air to reach the soil.

- **Gravel or Stones:** Ideal for decorative purposes and around perennial plants.

Benefits of Mulching

1. Moisture Conservation:
- Reduces soil evaporation, maintaining consistent moisture levels, which is crucial in dry and arid climates.

2. Weed Suppression:
- Blocks sunlight, preventing weed seeds from germinating, reducing the need for chemical herbicides and manual weeding.

3. Soil Temperature Regulation:
- Insulates the soil, keeping it cooler in summer and warmer in winter, protecting plant roots from extreme temperature fluctuations.

4. Soil Health Improvement:

- Organic mulches decompose, adding organic matter to the soil, enhancing soil structure, fertility, and microbial activity.

5. Erosion Control:
- Protects the soil surface from wind and water erosion, maintaining soil integrity and preventing nutrient runoff.

6. Pest and Disease Management:
- Some mulches, like cedar and pine, have natural pest-repellent properties and can reduce the spread of soil-borne diseases by preventing soil splashing onto plant leaves.

Mulching Techniques

1. Application:
- Apply mulch to a depth of 2-4 inches, depending on the material and garden needs, and keep mulch away from the base of plants to prevent rot and pest issues.

2. Timing:

- Apply mulch in spring after the soil has warmed and in fall to protect soil from winter weather, reapplying as needed to maintain the desired depth.

3. Selection:

- Choose mulch based on plant needs, climate, and garden aesthetics, using coarse mulch for pathways and fine mulch for garden beds.

Cover Crops: Principles and Practices

Cover crops are plants grown primarily to improve soil health rather than for harvest. They play a critical role in sustainable agriculture, enhancing soil fertility, structure, and biodiversity.

Types of Cover Crops

1. Legumes:

- **Examples**: Clover, vetch, peas, beans, alfalfa.
   - **Benefits**: Fix atmospheric nitrogen into the soil through symbiotic relationships with Rhizobium bacteria.

2. Grasses:
   - **Examples**: Rye, oats, barley, wheat.
   - Benefits: Provide biomass, improve soil structure, and suppress weeds.

3. Brassicas:
   - **Examples**: Mustard, radish, turnips.
   - Benefits: Break up compacted soil with their deep roots and suppress pests and diseases.

4. Broadleaf Plants:
   - **Examples**: Buckwheat, phacelia, sunflower.
   - **Benefits:** Attract beneficial insects and improve soil organic matter.

Benefits of Cover Crops

1. Soil Fertility and Nutrient Management:
   - Enhance soil nitrogen levels, especially with legumes, and deep-rooted cover crops mine nutrients from deeper soil layers, bringing them to the surface.

2. Soil Structure and Health:
   - Improve soil structure by increasing organic matter and enhancing microbial activity, with roots creating channels in the soil for better aeration and water infiltration.

3. Erosion Control:
   - Protect the soil surface from erosion caused by wind and water, with roots helping to bind the soil and reduce runoff and soil loss.

4. Weed Suppression:

- Dense growth shades out weeds, reducing their establishment and growth, with some cover crops releasing allelopathic chemicals that inhibit weed germination.

5. Pest and Disease Management:

- Disrupt pest and disease cycles by providing habitat for beneficial insects and predators, with some cover crops, like mustard, having biofumigant properties that suppress soil-borne pests and diseases.

6. Biodiversity Enhancement:

- Increase plant diversity in agricultural systems, promoting ecological balance and providing habitat for various beneficial insects, birds, and other wildlife.

Cover Crop Management

1. Selection:

- Choose cover crops based on soil needs, climate, and the cropping system, considering the benefits each cover crop provides and its compatibility with main crops.

2. Planting:
   - Plant during off-seasons or between main crop cycles, using appropriate seeding rates and methods (broadcasting, drilling) for optimal establishment.

3. Termination:
   - Terminate cover crops before they set seed to prevent them from becoming weeds, using methods such as mowing, tilling, crimping, or herbicides, depending on the farming system.

4. Incorporation:
   - Incorporate cover crop residues into the soil to decompose and release nutrients, with no-till systems using cover crops as mulch by leaving residues on the soil surface.

Integrating Mulching and Cover Crops

Combining mulching and cover cropping can significantly enhance soil health and crop productivity. Here are some strategies for integrating these practices:

1. Cover Crops as Mulch:
   - Terminate cover crops and use their residues as mulch for the next crop, providing soil protection, moisture conservation, and weed suppression.

2. Living Mulch:
   - Use low-growing cover crops as a living mulch under main crops, maintaining soil cover and benefiting soil health without competing excessively with main crops.

3. Sequential Planting:
   - Plant cover crops in rotation with main crops and use mulch during off-seasons to ensure continuous soil protection and nutrient cycling.

4. Complementary Benefits:
- Use mulches to retain soil moisture and cover crops to enhance soil fertility, creating a synergistic effect that improves overall soil health.

Challenges and Solutions

1. Cover Crop Management:
- **Challenge:** Timing cover crop planting and termination can be challenging.
- Solution: Plan cover crop cycles carefully and use appropriate termination methods.

2. Mulch Availability:
- **Challenge:** Securing a consistent supply of organic mulch can be difficult.
- **Solution:** Source locally available materials and consider growing your own mulch plants.

3. Weed Control:

- **Challenge**: Weeds can still emerge through mulch and cover crops.
   - **Solution:** Use thicker mulch layers and select competitive cover crop species.

4. Pest and Disease Management:
   - **Challenge:** Certain mulches and cover crops can harbor pests and diseases.
   - **Solution**: Rotate mulch materials and cover crop species to prevent the buildup of specific pests and diseases.

Conclusion

Mulching and cover cropping are foundational practices for sustainable gardening and farming. They enhance soil health, conserve resources, and improve crop productivity. By understanding the principles, types, benefits, and management of mulches and cover crops, gardeners and farmers can create resilient, productive, and sustainable agricultural systems. Integrating these practices ensures long-term soil health, ecological balance, and sustainable food production, supporting the overall health of the environment and the communities that depend on it.

2. Soil Testing and Amendments

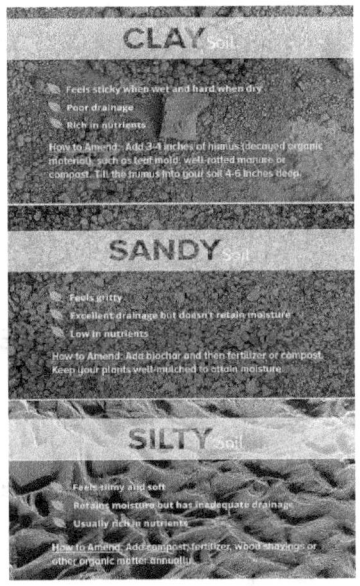

Soil Testing and Amendments: A Comprehensive Guide

Understanding and improving soil health is crucial for successful gardening and farming. Soil testing provides valuable insights into the soil's composition, fertility, and needs, while soil amendments help improve its structure, nutrient content, and overall health. This comprehensive guide explores the importance of soil testing, the different types of tests, how to interpret results, and the various soil amendments used to enhance soil quality.

Soil Testing: An Overview

Soil testing is the process of analyzing soil samples to determine their nutrient content, pH level, and other key characteristics. Regular soil testing helps gardeners and farmers make informed decisions about soil management and crop production.

Importance of Soil Testing

1. Nutrient Management:

- Determines the levels of essential nutrients in the soil, such as nitrogen (N), phosphorus (P), potassium (K), and trace elements.

- Helps identify nutrient deficiencies or excesses that can affect plant growth and health.

2. pH Assessment:

- Measures soil pH to determine its acidity or alkalinity, which affects nutrient availability and microbial activity.

- Helps guide the application of lime or sulfur to adjust soil pH to optimal levels for specific crops.

3. Soil Structure and Health:

- Provides insights into soil texture, organic matter content, and other physical properties.

- Helps identify issues like soil compaction, poor drainage, or low organic matter that can impact plant growth.

4. Environmental Protection:
 - Prevents over-fertilization and the associated environmental risks, such as nutrient runoff and water pollution.
 - Promotes sustainable soil management practices that protect natural resources.

Types of Soil Tests

1. Basic Soil Test:
 - Measures soil pH and levels of primary nutrients (N, P, K).
 - Suitable for general soil fertility assessment.

2. Comprehensive Soil Test:
 - Includes pH, primary nutrients, secondary nutrients (calcium, magnesium, sulfur), and micronutrients (iron, manganese, zinc, copper, boron).
 - Provides a detailed analysis of soil fertility and health.

3. Soil Texture Test:

- Determines the proportions of sand, silt, and clay in the soil.

- Helps assess soil drainage and aeration properties.

4. Organic Matter Test:

- Measures the percentage of organic matter in the soil.

- Indicates soil fertility, water retention, and microbial activity.

5. Salinity Test:

- Measures the soluble salt content in the soil.

- Important for managing soil health in areas with high salinity or irrigation issues.

6. Biological Soil Test:

- Assesses microbial activity and diversity in the soil.

- Provides insights into soil health and the effectiveness of organic matter management.

Soil Sampling Procedures

1. Timing:
- Conduct soil tests during the growing season or just before planting.
- Avoid testing immediately after fertilization or heavy rain.

2. Sampling Tools:
- Use clean tools, such as a soil probe, auger, or spade, to collect soil samples.
- Avoid using galvanized or brass tools that can contaminate samples with zinc or copper.

3. Sampling Depth:
- For lawns and gardens, sample to a depth of 6-8 inches.
- For trees and shrubs, sample to a depth of 12-18 inches.

4. Sampling Technique:
- Collect multiple subsamples from different locations within the testing area.
- Combine subsamples in a clean container to create a composite sample.

5. Sample Preparation:
- Remove debris, such as rocks and plant material, from the composite sample.
- Air-dry the soil sample and place it in a labeled bag for testing.

Interpreting Soil Test Results

1. Nutrient Levels:
- Compare nutrient levels to recommended ranges for specific crops.
- Identify deficiencies or excesses that require management.

2. Soil pH:

- Assess whether the soil pH is within the optimal range (typically 6.0-7.0) for the desired crops.

- Plan pH adjustments if needed using lime (to raise pH) or sulfur (to lower pH).

3. Organic Matter:

- Evaluate the percentage of organic matter in the soil.

- Aim for 3-5% organic matter in most garden soils.

4. Recommendations:

- Follow specific recommendations provided in the soil test report for fertilization and amendments.

- Adjust management practices based on test results to improve soil health and crop productivity.

Soil Amendments: Enhancing Soil Health

Soil amendments are materials added to the soil to improve its physical, chemical, and biological properties. They play a crucial role in enhancing soil structure, fertility, and overall health.

Types of Soil Amendments

1. Organic Amendments:

- **Compost:** Decomposed organic matter that enriches soil with nutrients and improves its structure and water retention.

- **Manure**: Animal waste that provides nutrients and organic matter; should be well-composted to avoid pathogens and weed seeds.

- **Green Manure**: Cover crops grown and then incorporated into the soil to add organic matter and nutrients.

- **Peat Moss**: Increases soil organic matter and water retention; typically used in acidic soils.

- **Worm Castings:** Rich in nutrients and beneficial microbes; improves soil structure and fertility.

2. Inorganic Amendments:
- **Lime**: Raises soil pH and supplies calcium and magnesium; used in acidic soils.
- **Gypsum**: Provides calcium and sulfur without altering soil pH; improves soil structure in saline or sodic soils.
- **Sand**: Improves drainage and aeration in clay soils; should be used with caution to avoid creating a concrete-like texture.
- **Perlite and Vermiculite**: Lightweight materials that enhance soil aeration and water retention; commonly used in potting mixes.

3. Mineral Amendments:
- Rock Phosphate: A slow-release source of phosphorus for plants; improves soil fertility.

- **Greensand:** Provides potassium, iron, and trace minerals; enhances soil structure and nutrient availability.
- **Azomite**: A natural mineral product rich in trace elements; enhances soil fertility and plant health.

Benefits of Soil Amendments

1. Improved Soil Structure:
- Organic amendments increase soil aggregation, improving porosity and water infiltration.
- Reduces soil compaction and enhances root growth and development.

2. Enhanced Nutrient Availability:
- Organic amendments provide a slow-release source of nutrients, improving soil fertility.

- Mineral amendments supply essential nutrients and trace elements required for plant growth.

3. Increased Water Retention:
- Organic amendments improve soil's ability to retain water, reducing irrigation needs.
- Beneficial in sandy soils that drain quickly and in regions with low rainfall.

4. Enhanced Microbial Activity:
- Organic amendments support a diverse and active soil microbial community.
- Improves nutrient cycling, disease suppression, and overall soil health.

5. pH Adjustment:
- Lime and sulfur are used to adjust soil pH to optimal levels for plant growth.
- Proper pH ensures nutrient availability and effective microbial activity.

Applying Soil Amendments

1. Incorporation:
- Incorporate amendments into the soil before planting or during soil preparation.
- Use tillage or hand tools to mix amendments evenly into the top 6-8 inches of soil.

2. Top-Dressing:
- Apply amendments as a surface layer around established plants.
- Ideal for compost, mulch, and slow-release fertilizers.

3. Mulching:
- Use organic materials like straw, wood chips, or compost as mulch.
- Protects soil from erosion, conserves moisture, and adds organic matter as it decomposes.

4. Composting:
- Compost organic materials to create nutrient-rich soil amendments.

- Apply finished compost to garden beds, lawns, and potting mixes.

5. Green Manure:
- Grow cover crops and incorporate them into the soil before they set seed.
- Improves soil organic matter and fertility.

Best Practices for Soil Amendment

1. Test Soil Regularly:
- Conduct soil tests to determine nutrient needs and appropriate amendments.
- Adjust amendment practices based on soil test results.

2. Use Quality Materials:
- Use high-quality, well-composted organic materials to avoid introducing pathogens or weed seeds.

- Source amendments from reputable suppliers.

3. Balance Amendments:
- Avoid over-application of amendments that can lead to nutrient imbalances or soil pH issues.
- Follow recommended application rates and guidelines.

4. Monitor Soil Health:
- Observe plant growth, soil structure, and microbial activity to assess the effectiveness of amendments.
- Adjust practices as needed to maintain optimal soil health.

5. Sustainable Practices:
- Use amendments that support sustainable soil management and environmental protection.

- Minimize reliance on synthetic fertilizers and pesticides.

Conclusion

Soil testing and amendments are essential components of sustainable gardening and farming. Regular soil testing provides critical information on soil health and fertility, guiding the effective use of soil amendments. By understanding the principles, types, benefits, and application methods of soil amendments, gardeners and farmers can enhance soil structure, nutrient availability, and overall soil health. Implementing these practices ensures long-term productivity, environmental protection, and sustainable land management, supporting healthy plant growth and resilient ecosystems.

Understanding Soil Types

Understanding Soil Types: A Comprehensive Guide

Understanding soil types is fundamental for effective gardening and farming. Different soil types have distinct physical, chemical, and biological properties that influence their behavior, fertility, and suitability for various crops. This guide delves into the characteristics of different soil types, how to identify them, and the best management practices for each.

Soil Composition and Texture

Soil is composed of minerals, organic matter, water, and air. The relative proportions of sand, silt, and clay particles determine the soil's texture, which significantly affects its properties and behavior.

Soil Particle Sizes

1. Sand:
   - Size: 0.05 to 2 mm in diameter.
   - Characteristics: Coarse and gritty, with large spaces between particles.
   - Properties: Good drainage and aeration, but poor water and nutrient retention.

2. Silt:
   - Size: 0.002 to 0.05 mm in diameter.

- Characteristics: Smooth and floury when dry, slightly plastic when wet.
   - Properties: Intermediate drainage and nutrient-holding capacity.

3. Clay:
   - Size: Less than 0.002 mm in diameter.
   - Characteristics: Very fine particles that stick together and feel sticky when wet.
   - Properties: Poor drainage, high water and nutrient retention, prone to compaction.

Soil Texture Classes

The USDA soil texture triangle categorizes soils into 12 classes based on the proportions of sand, silt, and clay:

1. Sandy Soils:
   - Types: Sand, loamy sand.
   - Characteristics: High sand content (over 70%), excellent drainage, low fertility.

2. Loamy Soils:
 - Types: Sandy loam, loam, silt loam.
 - Characteristics: Balanced mixture of sand, silt, and clay, good drainage, and fertility.

3. Clayey Soils:
 - Types: Silty clay, clay loam, clay.
 - Characteristics: High clay content (over 40%), poor drainage, high fertility.

4. Silty Soils:
 - Types: Silt, silty clay loam.
 - **Characteristics:** High silt content, moderate drainage, and fertility.

Identifying Soil Types

Soil type can be identified through a combination of field tests and laboratory analysis. Here are some common methods:

Field Tests

1. Feel Test:
   - Method: Moisten a small amount of soil and rub it between your fingers.
   - Sand: Feels gritty.
   - Silt: Feels smooth and floury.
   - Clay: Feels sticky and can be molded.

2. Ribbon Test:
   - Method: Moisten soil, roll it into a ball, and squeeze it into a ribbon between your thumb and forefinger.
   - Sand: Breaks apart quickly.
   - Silt: Forms a short ribbon (less than 1 inch).
   - Clay: Forms a long, flexible ribbon (over 2 inches).

3. Jar Test:
   - **Method**: Fill a jar with soil and water, shake vigorously, and let it settle for 24 hours.
   - Sand: Settles quickly at the bottom.

- Silt: Forms a middle layer.
- Clay: Settles last, forming the top layer.

Laboratory Analysis

1. Particle Size Analysis:
 - **Method**: Laboratory testing using sieves and sedimentation to determine the proportions of sand, silt, and clay.
 - **Results:** Provides precise soil texture classification.

2. Soil Profile Examination:
 - Method: Dig a soil pit or use a soil auger to examine soil layers (horizons) and their characteristics.
 - **Results:** Provides insights into soil structure, color, root depth, and drainage properties.

Characteristics of Different Soil Types

Each soil type has unique characteristics that influence its suitability for various crops and management practices.

Sandy Soils

Characteristics:
- **Texture**: Coarse, gritty.
- Drainage: Excellent, dries out quickly.
- Water Retention: Low, prone to drought.
- Nutrient Holding: Poor nutrients leach easily.
- Aeration: Excellent.

Management Practices:
- Watering: Frequent, light watering to prevent drought stress.
- Mulching: Use organic mulch to conserve moisture.

- **Organic Matter:** Add compost and organic matter to improve water and nutrient retention.
- **Cover Crops**: Plant cover crops to reduce erosion and improve soil structure.

Suitable Crops:
- Root vegetables (carrots, potatoes), drought-tolerant plants (lavender, rosemary).

Loamy Soils

Characteristics:
- Texture: Balanced, ideal mixture of sand, silt, and clay.
- Drainage: Good.
- Water Retention: Adequate.
- Nutrient Holding: Good.
- **Aeration: Balanced.**

Management Practices:
- Watering: Regular, deep watering to encourage deep root growth.

- Organic Matter: Maintain organic matter levels with compost and green manure.
- Crop Rotation: Practice crop rotation to maintain soil fertility and structure.

Suitable Crops:
- Most vegetables, fruits, flowers, and grains.

Clayey Soils

Characteristics:
- Texture: Fine, sticky.
- Drainage: Poor, prone to waterlogging.
- Water Retention: High.
- Nutrient Holding: High.
- Aeration: Poor.

Management Practices:
- **Drainage**: Improve drainage with raised beds or drainage tiles.
- **Organic Matter**: Add organic matter to improve soil structure and aeration.

- **Avoid Compaction:** Minimize tilling and foot traffic when soil is wet.
- **Cover Crops:** Use deep-rooted cover crops to improve soil structure.

Suitable Crops:
- Water-loving plants (rice, taro), heavy feeders (cabbage, broccoli).

Silty Soils

Characteristics:
- Texture: Smooth, floury.
- Drainage: Moderate.
- Water Retention: Good.
- Nutrient Holding: Good.
- Aeration: Moderate.

Management Practices:
- **Erosion Control**: Use mulches and cover crops to prevent erosion.
- **Organic Matter:** Add compost to improve structure and nutrient availability.

- **Watering**: Ensure proper irrigation to avoid waterlogging.

Suitable Crops:
- Most vegetables, grains, and fruits.

Improving Soil Types

While each soil type has its inherent characteristics, management practices can significantly improve their properties to suit specific gardening and farming needs.

Improving Sandy Soils

1. Add Organic Matter:
   - Incorporate compost, manure, or peat moss to increase water and nutrient retention.

2. Mulching:
   - Use organic mulch to conserve moisture and reduce temperature fluctuations.

3. Cover Crops:
- Plant cover crops like clover or rye to add organic matter and reduce erosion.

4. Frequent Watering:
- Implement a consistent watering schedule to prevent soil from drying out.

Improving Clayey Soils

1. Add Organic Matter:
- Incorporate compost, leaf mold, or well-rotted manure to improve drainage and aeration.

2. Avoid Compaction:
- Limit tillage and heavy foot traffic, especially when the soil is wet.

3. Raised Beds:
- Use raised beds to improve drainage and root development.

4. Cover Crops:
- Use deep-rooted cover crops like daikon radish to break up compacted soil.

Improving Silty Soils

1. Add Organic Matter:
- Incorporate compost and other organic materials to improve soil structure and drainage.

2. Erosion Control:
- Use mulch and cover crops to protect the soil from erosion.

3. Proper Irrigation:
- Ensure adequate but not excessive watering to maintain soil moisture balance.

4. Avoid Compaction:
- Minimize tillage and heavy equipment use to maintain soil structure.

Conclusion

Understanding soil types is crucial for successful gardening and farming. Each soil type has unique characteristics that influence its suitability for different crops and management practices. By identifying soil types through field tests and laboratory analysis, and implementing appropriate management practices, gardeners and farmers can improve soil health, enhance plant growth, and achieve sustainable land management. Through the addition of organic matter, proper watering, and the use of cover crops, even the most challenging soils can be transformed into fertile, productive ground.

Organic Soil Amendments

Organic Soil Amendments: A Comprehensive Guide

Organic soil amendments are essential components of sustainable gardening and agriculture. They play a crucial role in improving soil health, structure, fertility, and overall ecosystem function. This guide explores the types, benefits, application methods, and best practices for using organic soil amendments to enhance your garden or farm's productivity and sustainability.

What Are Organic Soil Amendments?

Organic soil amendments are natural materials added to the soil to improve its physical properties, nutrient content, and biological activity. Unlike synthetic fertilizers, these amendments release nutrients slowly, support beneficial soil organisms, and contribute to the long-term health and fertility of the soil.

Types of Organic Soil Amendments

Organic soil amendments come in various forms, each offering unique benefits to soil and plants. Here are some of the most common types:

1. Compost

Description:
- Compost is decomposed organic matter made from kitchen scraps, garden waste, leaves, and manure.

Benefits:
- Improves soil structure, aeration, and water retention.
- Adds essential nutrients and organic matter to the soil.
- Enhances microbial activity and supports a healthy soil ecosystem.

Application:
- Apply a 2-4 inch layer of compost to garden beds and mix it into the top 6-8 inches of soil.
- Use compost as a mulch around plants to conserve moisture and suppress weeds.

2. Manure

Description:
- Animal waste from cows, horses, chickens, and other livestock, often composted to reduce pathogens and improve safety.

Benefits:
- Provides a rich source of nitrogen and other nutrients.
- Enhances soil structure and water-holding capacity.
- Stimulates microbial activity and decomposition.

Application:
- Use well-composted manure to avoid the risk of pathogens and weed seeds.
- Apply a 1-2 inch layer to garden beds and incorporate it into the soil.
- Avoid using fresh manure directly on edible crops to prevent contamination.

3. Green Manure and Cover Crops

Description:
- Plants grown specifically to be incorporated into the soil, such as clover, rye, and vetch.

Benefits:
- Adds organic matter and nutrients to the soil.
- Improves soil structure and reduces erosion.
- Suppresses weeds and breaks pest and disease cycles.

Application:
- Sow cover crops in empty garden beds during the off-season.
- Mow or chop the plants before they set seed and incorporate them into the soil.

4. Peat Moss

Description:
- Decomposed sphagnum moss harvested from peat bogs, commonly used for its acidic properties.

Benefits:
- Increases soil organic matter and water retention.
- Improves soil aeration and structure.

Application:
- Mix peat moss into garden soil at a rate of 1-2 inches per 6 inches of soil depth.
- Use as a soil amendment for acid-loving plants like blueberries and azaleas.

5. Worm Castings

Description:
- The nutrient-rich waste produced by earthworms, also known as vermicompost.

Benefits:
- Provides a balanced source of nutrients and beneficial microorganisms.
- Improves soil structure and water-holding capacity.
- Enhances plant growth and resistance to pests and diseases.

Application:
- Apply worm castings as a top dressing or mix into the soil at a rate of 1-2 cups per plant.
- Use in potting mixes to improve nutrient content and soil health.

6. Leaf Mold

Description:
- Partially decomposed leaves collected from deciduous trees, valued for its organic matter content.

Benefits:
- Increases soil organic matter and water retention.
- Improves soil structure and aeration.
- Provides a slow-release source of nutrients.

Application:
- Collect leaves in the fall and allow them to decompose for 1-2 years.
- Apply a 1-2 inch layer of leaf mold to garden beds and mix it into the soil.

7. Biochar

Description:
- Charcoal produced from organic materials like wood, crop residues, and manure through a process called pyrolysis.

Benefits:
- Enhances soil fertility by retaining nutrients and water.
- Improves soil structure and supports beneficial microbial activity.
- Sequesters carbon, contributing to climate change mitigation.

Application:
- Mix biochar with compost or manure before applying to the soil.
- Apply at a rate of 5-10% by volume of soil.

8. Bone Meal

Description:
- Finely ground animal bones, rich in phosphorus and calcium.

Benefits:
- Provides a slow-release source of phosphorus, essential for root development and flowering.
- Supplies calcium, important for cell wall strength and overall plant health.

Application:
- Apply bone meal during planting by mixing it into the soil at a rate of 1-2 tablespoons per planting hole.
- Use as a top dressing around established plants to boost phosphorus levels.

9. Blood Meal

Description:
- Dried and powdered animal blood, high in nitrogen.

Benefits:
- Provides a quick-release source of nitrogen, promoting vigorous leafy growth.
- Enhances microbial activity and decomposition in the soil.

Application:
- Apply blood meal by mixing it into the soil at a rate of 1-2 tablespoons per square foot.
- Use as a side dressing during the growing season to boost nitrogen levels.

10. Seaweed and Kelp

Description:
- Dried and processed seaweed, rich in micronutrients and growth hormones.

Benefits:
- Supplies a wide range of micronutrients, including iodine, potassium, and trace elements.
- Stimulates plant growth and enhances stress resistance.

Application:
- Apply seaweed meal by mixing it into the soil at a rate of 1-2 pounds per 100 square feet.
- Use liquid seaweed extract as a foliar spray or soil drench according to label instructions.

Benefits of Organic Soil Amendments

Organic soil amendments offer numerous benefits that contribute to sustainable soil management and healthy plant growth:

Improved Soil Structure

- **Aggregation:** Organic matter helps bind soil particles into aggregates, enhancing soil structure and stability.
- **Porosity:** Improved soil structure increases porosity, allowing better air and water movement.
- **Compaction Resistance:** Organic matter reduces soil compaction, promoting root growth and microbial activity.

Enhanced Water Retention and Drainage

- **Water Holding:** Organic amendments increase the soil's ability to retain water, reducing the need for frequent irrigation.
- **Drainage:** Improved soil structure enhances drainage, preventing waterlogging and root rot.

Increased Nutrient Availability

- **Slow Release:** Organic amendments provide a slow-release source of nutrients, reducing the risk of nutrient leaching.
- **Microbial Activity:** Enhanced microbial activity promotes nutrient cycling and availability to plants.
- **Cation Exchange Capacity (CEC):** Organic matter increases soil CEC, improving the soil's ability to hold and exchange nutrients.

Enhanced Soil Biological Activity

- **Microbial Diversity:** Organic amendments support a diverse and active microbial community, essential for soil health.
- **Beneficial Organisms:** Increased microbial activity promotes the presence of beneficial organisms, such as earthworms and mycorrhizal fungi.

Improved Soil pH

- **Buffering Capacity**: Organic matter helps buffer soil pH, making it more stable and suitable for a wider range of plants.
- **pH Adjustment:** Some organic amendments, like peat moss, can help adjust soil pH for specific crops.

Reduced Soil Erosion

- **Soil Cover:** Organic amendments, especially cover crops and mulches, protect the soil surface from erosion by wind and water.
- **Improved Aggregation:** Enhanced soil structure reduces the susceptibility to erosion.

Climate Change Mitigation

- **Carbon Sequestration:** Organic amendments, particularly biochar, sequester

carbon in the soil, reducing greenhouse gas emissions.
- **Sustainable Practices:** Using organic amendments reduces reliance on synthetic fertilizers, promoting sustainable and environmentally friendly practices.

Application Methods

Applying organic soil amendments correctly is crucial to maximizing their benefits. Here are some common methods:

Incorporation

- **Method:** Mix organic amendments into the soil during soil preparation or planting.
- **Benefits:** Improves soil structure and nutrient availability in the root zone.
- **Technique**: Use a tiller, spade, or garden fork to incorporate amendments into the top 6-12 inches of soil.

Top-Dressing

- **Method:** Apply organic amendments on the soil surface around plants.
- **Benefits:** Provides nutrients and organic matter without disturbing the soil.
- **Technique:** Spread a thin layer (1-2 inches) of compost, manure, or worm castings around the base of plants.

Mulching

- **Method:** Use organic materials like straw, wood chips, or leaf mold as mulch.
- **Benefits:** Conserves moisture, suppresses weeds, and adds organic matter as it decomposes.
- **Technique**: Apply a 2-4 inch layer of mulch around plants, keeping it away from the stems to prevent rot.

Compost Tea

- **Method:** Brew compost in water to create a nutrient-rich liquid.
- **Benefits:** Provides a quick boost of nutrients and beneficial microbes to plants.
- **Technique**: Steep compost in water for 24-48 hours, strain, and apply the liquid to soil or as a foliar spray.

Cover Cropping

- **Method**: Grow cover crops and incorporate them into the soil before they set seed.
- **Benefits:** Adds organic matter, improves soil structure, and enhances nutrient availability.
- **Technique:** Sow cover crop seeds in the garden bed, allow them to grow, and then mow or chop them before incorporation.

Best Practices for Using Organic Soil Amendments

To effectively use organic soil amendments and maximize their benefits, consider the following best practices:

1. Soil Testing

Why:
- Soil testing helps determine nutrient deficiencies, pH levels, and overall soil health, providing a baseline for amendment needs.

How:
- Collect soil samples from different areas of your garden or field.
- Send samples to a local extension service or soil testing laboratory.
- Follow recommendations based on test results to select appropriate amendments.

2. Understanding Soil Type

Why:
- Different soil types (sand, silt, clay) have unique characteristics and respond differently to amendments.

How:
- Identify your soil type through a simple jar test or by consulting local soil surveys.
- Tailor amendments to address specific soil needs (e.g., adding organic matter to sandy soils to improve water retention).

3. Appropriate Amendment Selection

Why:
- Not all amendments are suitable for all soil types or crops.

How:
- Match amendments to specific soil deficiencies and crop requirements.

- Use compost and manure for general soil improvement, peat moss for acid-loving plants, and biochar for carbon sequestration.

4. Correct Application Rates

Why:
- Over-application can lead to nutrient imbalances, environmental pollution, and waste of resources.

How:
- Follow recommended application rates based on soil test results and amendment guidelines.
- Avoid applying excessive amounts of high-nitrogen amendments like blood meal or fresh manure.

5. Timing and Seasonality

Why:
- Timing affects nutrient availability and soil health.

How:
- Apply amendments in the fall or early spring for optimal integration into the soil.
- Use cover crops during the off-season to add organic matter and protect the soil.

6. Incorporation and Mixing

Why:
- Proper incorporation ensures even distribution of nutrients and organic matter.

How:
- Use tools like tillers, spades, or garden forks to mix amendments into the soil.
- Incorporate amendments to a depth of 6-12 inches, depending on soil and crop needs.

7. Monitoring and Adjustment

Why:
- Continuous monitoring helps track soil health and improve effectiveness.

How:
- Regularly test soil to monitor changes in nutrient levels and pH.
- Adjust amendment practices based on observed plant growth and soil test results.

8. Diversifying Amendments

Why:
- Using a variety of amendments can address multiple soil needs and support a balanced ecosystem.

How:
- Combine different amendments like compost, manure, and cover crops to enhance soil health.

- Rotate cover crops to break pest and disease cycles and improve soil fertility.

9. Mulching

Why:
- Mulching conserves moisture, suppresses weeds, and adds organic matter.

How:
- Apply a 2-4 inch layer of organic mulch (e.g., straw, wood chips) around plants.
- Refresh mulch annually to maintain effectiveness.

10. Sustainable Sourcing

Why:
- Sustainable sourcing minimizes environmental impact and supports ecological balance.

How:
- Use locally sourced, sustainable amendments whenever possible.

- Avoid amendments from over-exploited or non-renewable resources (e.g., peat moss).

11. Avoiding Contaminants

Why:
- Contaminants can harm soil health and plant growth.

How:
- Use well-composted materials to reduce pathogens and weed seeds.
- Ensure manure is aged or composted to minimize risks.

12. Education and Adaptation

Why:
- Continuous learning and adaptation improve soil management practices over time.

How:
- Stay informed about new research and best practices in organic soil management.
- Adapt practices based on your experiences and evolving soil conditions.

By following these best practices, gardeners and farmers can effectively use organic soil amendments to improve soil health, enhance plant growth, and promote sustainable agricultural practices.

CHAPTER 3

Water Management

Water Management in Permaculture Gardening

Effective water management is crucial in permaculture gardening to ensure sustainable and productive systems. Water is a finite resource, and managing it wisely is essential for maintaining soil health, supporting plant growth, and conserving the environment. This guide covers the principles, techniques, and best practices for efficient water management in permaculture gardening.

Principles of Water Management in Permaculture

Permaculture emphasizes working with natural systems to create self-sustaining ecosystems. The principles of water management in permaculture include:

1. Catch and Store Water

Objective:
- Capture and store water during periods of abundance (e.g., rainfall) for use during dry periods.

Techniques:
- Install rainwater harvesting systems, such as barrels, tanks, or cisterns.
- Design landscapes to capture and direct rainwater into the soil using swales, ponds, and other earthworks.

2. Slow, Spread, and Sink

Objective:
- Slow down the flow of water to maximize infiltration and reduce runoff.

Techniques:
- Use swales, terraces, and contour planting to slow water movement.
- Implement mulch and cover crops to improve soil structure and water infiltration.
- Design landscapes to spread water evenly and allow it to sink into the ground.

3. Use Water Efficiently

Objective:
- Minimize water use and waste through efficient irrigation and soil management practices.

Techniques:
- Implement drip irrigation systems to deliver water directly to plant roots.

- Use mulch to reduce evaporation and retain soil moisture.
- Select drought-tolerant and native plants that require less water.

4. Recycle and Reuse Water

Objective:
- Recycle and reuse water within the system to reduce demand on external water sources.

Techniques:
- Use greywater systems to recycle household water for irrigation.
- Design ponds and wetlands to treat and reuse water.
- Capture and reuse runoff from impervious surfaces like roofs and driveways.

Techniques for Water Management in Permaculture

Effective water management involves a combination of techniques tailored to specific site conditions. Here are some key techniques used in permaculture gardening:

1. Rainwater Harvesting

Description:
- Collecting and storing rainwater from roofs and other surfaces for later use.

Benefits:
- Reduces reliance on municipal water supplies.
- Provides a free source of high-quality water.
- Helps mitigate stormwater runoff and erosion.

Implementation:
- Install gutters and downspouts to direct water into storage containers.
- Use rain barrels, tanks, or cisterns to store collected water.
- Incorporate overflow systems to handle excess water and prevent flooding.

2. Swales

Description:
- Contour ditches or trenches designed to capture and infiltrate rainwater.

Benefits:
- Slows water runoff, allowing it to soak into the soil.
- Reduces erosion and improves soil moisture.
- Enhances groundwater recharge and supports plant growth.

Implementation:
- Dig swales along contour lines of the landscape.
- Plant water-loving plants or cover crops in and around swales to enhance water absorption.
- Use swales in combination with other earthworks, such as terraces and ponds.

3. Drip Irrigation

Description:
- A low-pressure, low-volume irrigation system that delivers water directly to plant roots.

Benefits:
- Reduces water waste through evaporation and runoff.
- Provides precise and efficient water delivery to plants.
- Improves plant health by minimizing water contact with leaves and stems.

Implementation:
- Install drip lines or emitters around plants or along rows.
- Connect the system to a water source, such as a rain barrel or tap.
- Use timers or moisture sensors to automate irrigation and optimize water use.

4. Mulching

Description:
- Applying a layer of organic or inorganic material on the soil surface.

Benefits:
- Reduces evaporation and retains soil moisture.
- Suppresses weed growth and improves soil structure.
- Provides insulation and protects soil from temperature fluctuations.

Implementation:
- Apply a 2-4 inch layer of mulch around plants and garden beds.

- Use organic materials such as straw, wood chips, leaves, or compost.
- Replenish mulch annually or as needed to maintain effectiveness.

5. Cover Cropping

Description:
- Growing specific crops to cover the soil and provide various benefits.

Benefits:
- Improves soil structure and moisture retention.
- Reduces erosion and runoff.
- Adds organic matter and nutrients to the soil.

Implementation:
- Select appropriate cover crops based on climate and soil conditions (e.g., clover, rye, vetch).
- Sow cover crops during fallow periods or as part of crop rotation.

- Incorporate cover crops into the soil before planting the next crop.

6. Greywater Systems

Description:
- Recycling household water from sinks, showers, and laundry for irrigation.

Benefits:
- Reduces demand on potable water supplies.
- Provides a sustainable source of irrigation water.
- Reuses water that would otherwise be wasted.

Implementation:
- Install greywater diversion systems to collect and filter household water.
- Use greywater-friendly soaps and detergents to prevent soil contamination.
- Design irrigation systems to distribute greywater evenly and avoid overwatering.

7. Ponds and Wetlands

Description:
- Creating water features to capture, store, and recycle water within the landscape.

Benefits:
- Provides habitat for wildlife and enhances biodiversity.
- Supports aquaculture and integrated water management.
- Acts as a natural filtration system to improve water quality.

Implementation:
- Design ponds and wetlands to fit the landscape and capture runoff.
- Use appropriate plant species to enhance water filtration and biodiversity.
- Integrate ponds with other water management systems, such as swales and greywater systems.

Best Practices for Water Management in Permaculture

To maximize the benefits of water management techniques, follow these best practices:

1. Site Assessment

Why:
- Understanding the site conditions is essential for designing effective water management systems.

How:
- Conduct a thorough site assessment to identify slopes, soil types, and existing water flow patterns.
- Use tools like topographic maps, soil tests, and climate data to inform your design.

2. Integrated Design

Why:
- Integrating water management systems with other permaculture elements creates a cohesive and efficient landscape.

How:
- Incorporate water management features into your overall permaculture design.
- Connect rainwater harvesting, swales, ponds, and greywater systems to maximize water use efficiency.

3. Diverse Planting

Why:
- Diverse plantings enhance soil health, water retention, and ecosystem resilience.

How:
- Use a mix of native, drought-tolerant, and water-loving plants to create a balanced ecosystem.

- Implement polycultures and companion planting to support diverse root structures and water use patterns.

4. Regular Maintenance

Why:
- Regular maintenance ensures the continued effectiveness of water management systems.

How:
- Monitor and maintain rainwater harvesting systems, swales, and irrigation systems regularly.
- Check for leaks, clogs, and other issues that could reduce efficiency.

5. Adaptive Management

Why:
- Adapting to changing conditions and new information improves long-term water management success.

How:
- Continuously monitor soil moisture, plant health, and water usage.
- Adjust practices based on observations, weather patterns, and soil conditions.

6. Education and Community Involvement

Why:
- Educating yourself and others about water management promotes sustainable practices and community resilience.

How:
- Stay informed about new techniques and best practices in permaculture water management.
- Share knowledge and collaborate with neighbors, community groups, and local organizations.

By implementing these principles, techniques, and best practices, permaculture gardeners can manage water efficiently and sustainably, ensuring the health and productivity of their gardens while conserving this precious resource.

1. Harvesting and Conserving Water

Harvesting and Conserving Water: A Comprehensive Guide

Water harvesting and conservation are essential practices in permaculture gardening to ensure a sustainable and resilient system. These practices help capture, store, and use water efficiently, reducing reliance on external water sources and promoting environmental stewardship. This guide explores the techniques, benefits, and best practices for harvesting and conserving water in permaculture gardens.

Understanding Water Harvesting

Water harvesting involves capturing and storing rainwater and runoff for later use. This practice can be implemented on various scales, from small household systems to large landscape designs.

Types of Water Harvesting Systems

1. Rainwater Harvesting
   - **Description**: Capturing and storing rainwater from rooftops and other surfaces.
   - **Components**: Includes gutters, downspouts, filters, storage tanks, and distribution systems.

- **Benefits:** Provides a free, high-quality water source; reduces demand on municipal water supplies; mitigates stormwater runoff.

2. Groundwater Recharge
- **Description:** Methods to increase groundwater levels by directing surface water into the soil.
- **Components**: Includes infiltration basins, trenches, and recharge wells.
- **Benefits:** Enhances groundwater supplies; reduces surface runoff and erosion; supports sustainable water management.

3. Surface Water Harvesting
- **Description:** Capturing and storing surface runoff in ponds, lakes, and reservoirs.
- **Components**: Includes dams, weirs, channels, and ponds.
- **Benefits:** Provides a large-scale water storage solution; supports irrigation,

livestock, and aquaculture; enhances biodiversity and habitat creation.

4. Greywater Systems

- **Description**: Recycling water from household sinks, showers, and laundry for irrigation.

- **Components:** Includes diversion systems, filters, storage tanks, and distribution networks.

- **Benefits:** Reduces potable water use; provides a sustainable source of irrigation water; reuses water that would otherwise be wasted.

Techniques for Harvesting Water

1. Rainwater Harvesting Systems

Design and Implementation:
- **Roof Collection**: Install gutters and downspouts to channel rainwater from roofs into storage tanks.

- **First-Flush Diverters:** Use devices to divert the initial flow of dirty rainwater away from storage tanks.
- **Storage Tanks:** Choose tanks made of food-grade materials to avoid contamination; position tanks to maximize gravity-fed distribution.
- **Filtration:** Implement filters to remove debris and contaminants before water enters the storage tank.

Usage:
- Use harvested rainwater for irrigation, livestock, and household non-potable uses.
- Connect storage tanks to drip irrigation systems for efficient water use.

2. Swales and Contour Bunds

Design and Implementation:
- **Swales:** Dig shallow trenches along contour lines to capture and infiltrate rainwater.

- **Contour Bunds:** Build raised barriers along contour lines to slow down and capture runoff.
- **Planting:** Establish vegetation in and around swales and bunds to enhance water infiltration and soil stability.

Usage:
- Use swales and contour bunds to recharge groundwater, reduce erosion, and support plant growth.
- Integrate these features into the landscape design to create a self-sustaining water management system.

3. Infiltration Basins and Trenches

Design and Implementation:
- **Infiltration Basins**: Create shallow depressions to capture and hold rainwater, allowing it to infiltrate into the soil.
- **Infiltration Trenches:** Dig narrow, deep trenches filled with gravel or other

permeable materials to facilitate water infiltration.

Usage:
- Use basins and trenches to manage stormwater, recharge groundwater, and reduce surface runoff.
- Position these features strategically to capture runoff from impervious surfaces like driveways and parking areas.

4. Ponds and Reservoirs

Design and Implementation:
- **Site Selection:** Choose locations with suitable soil and topography to minimize seepage and maximize water retention.
- **Construction:** Build dams or embankments using locally available materials; install overflow structures to manage excess water.
- **Lining:** Use natural clay or synthetic liners to prevent seepage and enhance water retention.

Usage:
- Use ponds and reservoirs for irrigation, livestock, aquaculture, and wildlife habitat.
- Integrate these water bodies into the landscape design to support biodiversity and ecosystem services.

Techniques for Conserving Water

1. Efficient Irrigation Systems

Drip Irrigation:
- Design and Implementation: Install drip lines or emitters around plants or along rows; connect to a water source like a rain barrel or tank.
- **Benefits:** Reduces water waste through evaporation and runoff; delivers water directly to plant roots; improves plant health and water use efficiency.

Soaker Hoses:
- Design and Implementation: Lay soaker hoses on the soil surface around plants;

connect to a water source with a timer or manual control.
- Benefits: Provides a slow, steady water supply; minimizes evaporation and runoff; enhances soil moisture and plant health.

2. Mulching

Materials:
- **Organic Mulch**: Use straw, wood chips, leaves, compost, or grass clippings.
- **Inorganic Mulch:** Use gravel, pebbles, or landscape fabric.

Application:
- Apply a 2-4 inch layer of mulch around plants and garden beds; avoid direct contact with plant stems to prevent rot.
- Replenish mulch annually or as needed to maintain effectiveness.

Benefits:
- Reduces evaporation and retains soil moisture.

- Suppresses weed growth and improves soil structure.
- Provides insulation and protects soil from temperature fluctuations.

3. Soil Management

Techniques:
- **Organic Matter Addition**: Incorporate compost, manure, and other organic materials to improve soil structure and water-holding capacity.
- **Cover Cropping:** Grow cover crops to protect soil, enhance moisture retention, and add organic matter.
- **No-Till Farming**: Minimize soil disturbance to maintain soil structure and moisture levels.

Benefits:
- Enhances soil health and water infiltration.
- Reduces erosion and runoff.
- Improves plant growth and resilience.

4. Plant Selection and Placement

Drought-Tolerant Plants:
- Choose native and adapted species that require less water.
- Group plants with similar water needs together to optimize irrigation.

Placement:
- Position plants strategically based on water availability and microclimates.
- Use windbreaks, shade structures, and companion planting to create favorable growing conditions.

Benefits:
- Reduces water demand and enhances plant health.
- Supports biodiversity and ecosystem resilience.

5. Greywater Recycling

Design and Implementation:

- Install greywater diversion systems to collect water from sinks, showers, and laundry.
- Use filters to remove contaminants and ensure safe water for irrigation.
- Design irrigation systems to distribute greywater evenly and avoid overwatering.

Usage:
- Use greywater for landscape irrigation, reducing demand on potable water supplies.
- Follow local regulations and guidelines to ensure safe and effective use.

Benefits of Water Harvesting and Conservation

1. Sustainable Water Use
- Reduces reliance on external water sources and municipal supplies.
- Promotes efficient and responsible water use in gardening and agriculture.

2. Environmental Protection
- Mitigates stormwater runoff and reduces erosion.
- Enhances groundwater recharge and supports ecosystem health.

3. Cost Savings
- Reduces water bills and irrigation costs.
- Minimizes the need for expensive water infrastructure and treatment.

4. Resilience and Food Security
- Ensures a reliable water supply during droughts and dry periods.
- Supports sustainable food production and increases self-sufficiency.

5. Climate Change Mitigation
- Reduces the carbon footprint associated with water transport and treatment.
- Promotes sustainable land and water management practices.

Best Practices for Water Harvesting and Conservation

1. Holistic Design

Integration:
- Design water harvesting and conservation systems as part of an integrated permaculture plan.
- Consider site-specific factors like topography, soil type, climate, and existing vegetation.

Synergy:
- Create synergies between water management, soil health, and plant selection to optimize resource use and ecosystem function.

2. Regular Maintenance

Inspection:
- Regularly inspect and maintain water harvesting systems, including gutters, filters, storage tanks, and distribution networks.
- Check for leaks, clogs, and other issues that could reduce efficiency.

Upkeep:
- Clean gutters, downspouts, and filters to ensure proper water flow and quality.
- Repair and replace components as needed to maintain system functionality.

3. Community Engagement

Education:
- Educate yourself and others about the importance of water harvesting and conservation.
- Share knowledge and collaborate with neighbors, community groups, and local organizations.

Involvement:
- Participate in community projects and initiatives to promote sustainable water management.
- Advocate for policies and practices that support water conservation and environmental stewardship.

4. Adaptive Management

Monitoring:
- Continuously monitor soil moisture, plant health, and water usage.
- Adjust practices based on observations, weather patterns, and soil conditions.

Flexibility:
- Be flexible and willing to adapt to changing conditions and new information.
- Stay informed about new techniques and best practices in water management.

By implementing these principles, techniques, and best practices, permaculture gardeners can effectively harvest and conserve water, ensuring the sustainability and resilience of their gardens while conserving this vital resource.

Rainwater Harvesting

Rainwater Harvesting: A Comprehensive Guide

Rainwater harvesting is a sustainable practice that involves capturing, storing, and using rainwater for various purposes. This technique is essential in permaculture gardening as it reduces reliance on municipal water supplies, mitigates stormwater runoff, and promotes environmental stewardship. This guide explores the principles, systems, benefits, and best practices for effective rainwater harvesting.

Principles of Rainwater Harvesting

Rainwater harvesting operates on the principle of collecting rainwater from surfaces where it falls and storing it for future use. The key principles include:

1. **Catchment Surface:** Typically, rooftops serve as the primary catchment surface.
2. **Conveyance System:** Gutters and downspouts direct collected rainwater into storage.
3. **Storage:** Tanks, barrels, or cisterns store the harvested rainwater.
4. **Distribution**: Pumps or gravity-fed systems distribute the stored water to where it's needed.

Types of Rainwater Harvesting Systems

1. Rooftop Rainwater Harvesting

Description: Capturing rainwater from the roof and directing it to storage tanks.

Components:
- **Catchment Area:** The roof surface where rainwater is collected.
- **Conveyance System**: Gutters and downspouts that transport water to storage tanks.
- **First-Flush Diverters:** Devices that divert the initial flow of water, which may contain debris and contaminants.
- **Storage Tanks**: Containers that store the harvested rainwater for future use.
- **Filtration Systems:** Filters that remove debris and contaminants before water enters the storage tanks.

2. Surface Water Harvesting

Description: Capturing rainwater from land surfaces using structures like ponds, reservoirs, and infiltration basins.

Components:
- **Catchment Area**: Land surfaces such as gardens, fields, or paved areas.

- **Conveyance System**: Channels, swales, or berms that direct water to storage or infiltration areas.
- **Storage Structures**: Ponds, reservoirs, or infiltration basins that hold or absorb the water.
- **Overflow Systems**: Structures that manage excess water to prevent flooding.

Benefits of Rainwater Harvesting

1. **Water Conservation**: Reduces dependence on municipal water supplies and groundwater.
2. **Cost Savings:** Lowers water bills and irrigation costs.
3. **Environmental Protection:** Mitigates stormwater runoff, reducing erosion and water pollution.
4. **Resilience:** Provides a reliable water source during droughts or water shortages.
5. **Sustainable Gardening:** Supports permaculture principles by creating self-sustaining systems.

Designing and Implementing a Rainwater Harvesting System

1. Site Assessment

Roof Area Calculation:
- Determine the potential catchment area by measuring the roof surface.
- Use the formula: Catchment Area (sq. ft.) = Roof Length × Roof Width.

Rainfall Data:
- Obtain local rainfall data to estimate the volume of water that can be harvested.
- Use the formula: Harvested Water Volume (gallons) = Catchment Area (sq. ft.) × Rainfall Depth (inches) × 0.623 (conversion factor).

2. System Components

Catchment Surface:
- Ensure the roof is made of non-toxic materials.

- Clean the roof regularly to remove debris and contaminants.

Gutters and Downspouts:
- Install gutters along the roof edge to capture rainwater.
- Use downspouts to direct water to storage tanks.

First-Flush Diverters:
- Install devices to divert the initial flow of water, which may contain debris and contaminants.
- Ensure the diverter can handle the expected volume of initial runoff.

Storage Tanks:
- Choose tanks made of food-grade materials to avoid contamination.
- Size tanks based on estimated water needs and available catchment area.

Filtration Systems:
- Install pre-storage filters to remove large debris and contaminants.
- Use fine filters or UV systems for potable water applications.

3. Water Distribution

Gravity-Fed Systems:
- Position storage tanks at higher elevations to use gravity for water distribution.
- Use hoses or pipes to direct water to garden beds or irrigation systems.

Pumped Systems:
- Install pumps to distribute water from storage tanks to desired locations.
- Use pressure regulators to maintain consistent water flow.

4. Maintenance

Regular Inspection:
- Inspect gutters, downspouts, and storage tanks for debris and damage.
- Clean filters and first-flush diverters to ensure efficient operation.

Tank Cleaning:
- Periodically clean storage tanks to remove sediment and prevent algae growth.
- Use non-toxic cleaning agents to avoid contaminating stored water.

System Upgrades:
- Upgrade components as needed to improve efficiency and capacity.
- Consider adding sensors or automation for easier management.

Best Practices for Rainwater Harvesting

1. Optimize Catchment Area

Why: Maximizes the volume of harvested water.

How:
- Clean and maintain roof surfaces regularly.
- Use larger roofs or multiple catchment surfaces if possible.

2. Ensure Water Quality

Why: Prevents contamination and ensures safe water use.

How:
- Install and maintain first-flush diverters and filtration systems.
- Use food-grade materials for all components in contact with water.

3. Match Storage to Needs

Why: Balances water supply with demand.

How:
- Calculate water needs based on garden size and plant requirements.
- Choose storage tanks that provide adequate capacity for anticipated water use.

4. Design for Overflow Management

Why: Prevents flooding and erosion.

How:
- Install overflow systems to direct excess water to safe areas.
- Use swales, ponds, or infiltration basins to manage overflow.

5. Integrate with Landscape Design

Why: Enhances efficiency and aesthetics.

How:
- Position storage tanks and distribution systems for easy access and minimal visual impact.
- Use harvested rainwater for multiple purposes, such as irrigation, livestock, and aquaculture.

6. Educate and Involve Community

Why: Promotes wider adoption and collective benefits.

How:
- Share knowledge and resources with neighbors and community groups.
- Participate in community rainwater harvesting projects and initiatives.

Advanced Techniques in Rainwater Harvesting

1. Automated Systems

Description: Use sensors and controllers to automate water collection, storage, and distribution.

Benefits:
- Enhances efficiency and convenience.
- Provides real-time monitoring and management.

Implementation:
- Install sensors to monitor tank levels and water quality.
- Use controllers to automate pumps and irrigation systems based on soil moisture and weather conditions.

2. Integrated Systems

Description: Combine rainwater harvesting with other sustainable practices, such as greywater recycling and solar power.

Benefits:
- Maximizes resource efficiency and sustainability.
- Supports a holistic approach to water and energy management.

Implementation:
- Design systems that integrate rainwater harvesting with greywater reuse for irrigation.
- Use solar panels to power pumps and automation systems.

3. Community-Based Harvesting

Description: Implement rainwater harvesting at a community scale to provide shared benefits.

Benefits:
- Enhances water security and resilience for the community.
- Reduces infrastructure costs through shared resources.

Implementation:
- Develop community gardens or green spaces with shared rainwater harvesting systems.
- Collaborate with local governments and organizations to fund and manage projects.

Case Studies and Examples

Urban Rooftop Harvesting:
- **Location**: A city apartment building.
- **System:** Rooftop catchment, first-flush diverters, storage tanks, and drip irrigation.
- **Impact:** Reduced water bills, improved garden health, and community engagement.

Rural Homestead Harvesting:
- **Location**: A rural permaculture farm.
- **System:** Roof catchment, swales, ponds, and gravity-fed irrigation.
- **Impact:** Enhanced water security, reduced erosion, and sustainable food production.

Community Rainwater Harvesting:
- **Location:** A suburban community garden.
- **System:** Combined rainwater and greywater harvesting, shared storage tanks, and automated irrigation.
- **Impact:** Increased water availability, reduced municipal water use, and community cohesion.

By understanding and implementing rainwater harvesting principles, systems, and best practices, permaculture gardeners can create sustainable, resilient, and productive landscapes. Rainwater harvesting not only conserves water but also enhances soil health, plant growth, and overall ecosystem balance. Through careful design, regular maintenance, and community involvement, rainwater harvesting can be a key component of sustainable gardening and water management.

Greywater Systems

Greywater Systems

Greywater systems are a vital component of sustainable water management in permaculture gardening. They involve the recycling of water from household sources such as sinks, showers, and laundry (excluding toilets) for irrigation and other non-potable uses. By reusing greywater, permaculture practitioners can reduce water consumption, minimize wastewater, and create a more resilient and self-sustaining system. This guide provides an extensive overview of greywater systems, including their benefits, design principles, components, implementation, and best practices.

What is Greywater?

Greywater is relatively clean wastewater generated from domestic activities such as bathing, hand washing, laundry, and dishwashing. It differs from blackwater, which contains waste from toilets and kitchen sinks and requires more intensive treatment.

Characteristics of Greywater:
- **Contains Low Levels of Contaminants:** Greywater may contain soap, detergents, food particles, hair, and skin cells.
- **Biodegradable:** Most contaminants in greywater are organic and can be broken down naturally.
- **Nutrient-Rich:** Greywater can contain beneficial nutrients like nitrogen and phosphorus, which can enhance soil fertility when used for irrigation.

Benefits of Greywater Systems

1. **Water Conservation:** Significantly reduces the demand for potable water by recycling household water.

2. Cost Savings: Lowers water bills and reduces the need for expensive irrigation systems.

3. Environmental Protection: Reduces the volume of wastewater entering septic systems and treatment plants, decreasing environmental pollution.

4. *Enhanced Soil Fertility*: Provides plants with additional nutrients, reducing the need for chemical fertilizers.

5. Increased Resilience: Ensures a reliable water supply for irrigation, especially in arid or drought-prone regions.

Design Principles of Greywater Systems

1. Simplicity: Design systems that are easy to install, use, and maintain.

2. Safety: Ensure that greywater systems do not pose health risks to users or the environment.

3. Efficiency: Maximize the reuse of greywater while minimizing water loss and energy consumption.

4. Scalability: Design systems that can be scaled up or down based on water needs and available resources.

5. Integration: Incorporate greywater systems into the overall permaculture design to enhance sustainability and resource efficiency.

Components of Greywater Systems

1. Source: The origin of greywater, typically sinks, showers, and laundry machines.

2. Diversion System: Mechanisms to redirect greywater from household plumbing to the treatment or storage system.

3. Filtration System: Filters to remove larger particles and contaminants from the greywater.

4. Treatment System: Biological or chemical processes to further clean the greywater, if necessary.

5. Storage System: Tanks or containers to hold treated greywater until it is needed.

6. Distribution System: Pipes, hoses, or drip irrigation systems to deliver greywater to plants or landscape areas.

Types of Greywater Systems

1. Direct Greywater Systems
- **Description**: Greywater is used immediately after diversion, with minimal treatment.
- **Components:** Diversion valves, filters, and irrigation systems.
- **Advantages:** Simple, low-cost, and effective for immediate use.
- **Disadvantages**: Limited storage capacity; potential for clogging if not properly filtered.

2. Gravity-Fed Systems
- **Description**: Uses gravity to distribute greywater from the source to the irrigation area.
- **Components:** Gravity-fed piping, distribution basins, and mulch basins.

- **Advantages:** Energy-efficient and low-maintenance.

- **Disadvantages**: Requires careful design to ensure proper flow and avoid waterlogging.

3. Pumped Systems
- **Description:** Uses pumps to move greywater from the source to the treatment or storage system and then to the irrigation area.

- **Components:** Pumps, filters, storage tanks, and distribution systems.

- **Advantages:** Flexible and can be used in flat or complex terrains.

- **Disadvantages:** Requires energy and regular maintenance.

4. Constructed Wetlands
- **Description:** Uses natural wetland plants and microorganisms to treat greywater in a designed wetland environment.

- **Components:** Wetland beds, plants, gravel, and liners.
- **Advantages:** Provides effective natural treatment and creates wildlife habitat.
- **Disadvantages**: Requires space and careful design to ensure proper function.

Designing and Implementing a Greywater System

1. Assessing Greywater Sources and Demand

Source Identification:
- Identify potential sources of greywater in the household (e.g., sinks, showers, laundry).
- Estimate the volume of greywater generated daily from each source.

Water Demand:
- Calculate the water needs of the garden or landscape area.

- Ensure that the greywater supply matches or exceeds the irrigation demand.

2. System Design and Layout

Diversion System:
- Install diversion valves to redirect greywater from household plumbing to the greywater system.
- Ensure that the diversion system is easily accessible and operable.

Filtration and Treatment:
- Install coarse filters to remove large particles and prevent clogging.
- Consider additional treatment methods (e.g., sand filters, biofilters) if using greywater for sensitive plants or close to edible crops.

Storage and Distribution:
- Choose storage tanks or containers based on the volume of greywater and irrigation needs.

- Design the distribution system to deliver greywater efficiently to plants, avoiding overwatering or waterlogging.

3. System Installation

Plumbing:
- Connect household plumbing to the greywater system using appropriate piping and fittings.
- Ensure that all connections are secure and leak-free.

Filtration:
- Install filters at key points to remove debris and contaminants from the greywater.
- Regularly clean and maintain filters to ensure proper function.

Storage:
- Position storage tanks in shaded areas to prevent algae growth.

- Install overflow mechanisms to handle excess greywater during periods of low demand.

Distribution:
- Use drip irrigation or sub-surface irrigation systems to distribute greywater evenly.
- Avoid using sprinklers or surface irrigation to minimize the risk of contact with greywater.

4. Maintenance and Monitoring

Regular Inspection:
- Inspect the entire greywater system regularly for leaks, clogs, and other issues.
- Ensure that all components are functioning properly and efficiently.

Filter Cleaning:
- Clean filters regularly to prevent clogging and maintain water flow.
- Replace filters as needed to ensure effective filtration.

System Adjustments:
- Monitor soil moisture and plant health to adjust greywater distribution as needed.
- Modify the system layout or components if necessary to improve performance.

Best Practices for Greywater Systems

1. Use Appropriate Detergents and Soaps

Why: Prevents harmful chemicals from entering the greywater system.

How:
- Use biodegradable, low-phosphate, and low-sodium detergents and soaps.
- Avoid products with harmful chemicals or synthetic fragrances.

2. Avoid Contaminants

Why: Ensures the safety and health of plants and soil.

How:
- Prevent greywater from containing hazardous substances like bleach, paint, or harsh chemicals.
- Educate household members about what can and cannot go down the drain.

3. Design for Safety and Compliance

Why: Ensures the system is safe and meets local regulations.

How:
- Follow local building codes and regulations for greywater systems.
- Install backflow preventers and air gaps to protect potable water supplies.

4. Monitor and Maintain System Health

Why: Ensures long-term functionality and efficiency.

How:
- Regularly inspect and maintain all components of the greywater system.
- Address any issues promptly to prevent system failure or water contamination.

5. Educate and Involve Users

Why: Promotes responsible use and maintenance.

How:
- Educate household members about the greywater system and its benefits.
- Encourage participation in system maintenance and monitoring.

6. Integrate with Permaculture Design

Why: Enhances overall system sustainability and resilience.

How:
- Integrate greywater systems with other permaculture elements like rainwater harvesting and composting.
- Design the landscape to maximize water use efficiency and support diverse plantings.

Case Studies and Examples

Urban Greywater Recycling:
- **Location:** A city apartment with a small balcony garden.
- **System:** Direct greywater system using kitchen sink water for container plants.
- **Impact:** Reduced water bills, healthier plants, and increased awareness of sustainable practices.

Rural Greywater System:
- **Location**: A rural homestead with extensive gardens.
- **System:** Gravity-fed greywater system using bathroom and laundry water for garden irrigation.
- **Impact:** Enhanced water security, improved soil fertility, and sustainable food production.

Community Greywater Project:
- **Location**: A suburban community garden.
- **System**: Constructed wetland to treat and reuse greywater from communal facilities.
- **Impact**: Increased water availability, improved community engagement, and enhanced biodiversity.

By understanding and implementing greywater systems, permaculture gardeners can significantly reduce their water footprint, enhance soil fertility, and create a more resilient and sustainable gardening system. With careful design, regular maintenance, and community involvement, greywater systems can be a valuable asset in sustainable water management.

2. Irrigation Techniques

Irrigation Techniques: A Comprehensive Guide

Irrigation is a critical aspect of gardening and agriculture, particularly in permaculture systems where efficient water use is essential for sustainability. Effective irrigation techniques ensure that plants receive the right amount of water at the right time, promoting healthy growth and productivity while conserving water resources. This guide covers a range of irrigation techniques, including their benefits, applications, design principles, and best practices.

Overview of Irrigation Techniques

Irrigation techniques can be broadly categorized into surface irrigation, sprinkler irrigation, and drip or micro-irrigation. Each method has its own advantages and is suitable for different types of crops, soil conditions, and landscapes.

Surface Irrigation

1. Furrow Irrigation

Description: Water is directed into shallow furrows or trenches between crop rows.

Applications: Suitable for row crops like corn, beans, and potatoes.

Advantages:
- Simple and cost-effective.
- Requires minimal infrastructure.

Disadvantages:
- Can lead to uneven water distribution.
- High water loss due to evaporation and runoff.

Best Practices:
- Level the land to ensure uniform water distribution.
- Use mulching to reduce evaporation and runoff.

2. Basin Irrigation

Description: Water is applied to flat, rectangular, or circular basins, typically used for orchard trees or paddy fields.

Applications: Suitable for orchards, rice paddies, and vegetable gardens.

Advantages:
- Efficient water use with minimal runoff.
- Provides deep water penetration to root zones.

Disadvantages:
- Requires precise leveling and basin construction.
- Labor-intensive setup and maintenance.

Best Practices:
- Regularly maintain basins to prevent leaks and erosion.

- Combine with mulching to conserve moisture.

3. Border Irrigation

Description: Water is applied to long, narrow strips bordered by raised earth embankments.

Applications: Suitable for pastures, forage crops, and grains.

Advantages:
- Efficient for large areas with uniform slopes.
- Simple design and low cost.

Disadvantages:
- Requires precise leveling.
- Potential for waterlogging and uneven distribution.

Best Practices:
- Ensure borders are well-maintained and leveled.
- Monitor soil moisture to prevent over-irrigation.

Sprinkler Irrigation

1. Overhead Sprinklers

Description: Water is sprayed over plants from overhead sprinklers, simulating natural rainfall.

Applications: Suitable for lawns, gardens, and field crops.

Advantages:
- Even water distribution.
- Can cover large areas efficiently.

Disadvantages:
- High initial cost for equipment.

- Water loss due to evaporation and wind drift.

Best Practices:
- Use in the early morning or late evening to reduce evaporation.
- Regularly check and maintain sprinkler heads for optimal performance.

2. Traveling Sprinklers

Description: A mobile sprinkler system moves across the field, spraying water uniformly.

Applications: Suitable for large fields and pastures.

Advantages:
- Can irrigate large areas with minimal labor.
- Adjustable speed and water application rates.

Disadvantages:
- Expensive equipment and maintenance.
- Requires a reliable power source.

Best Practices:
- Schedule irrigation during low wind periods to minimize water loss.
- Regularly inspect and service the sprinkler system.

3. Center Pivot Irrigation

Description: A central pivot point with rotating sprinkler arms irrigates circular fields.

Applications: Suitable for large-scale agricultural fields.

Advantages:
- Efficient and uniform water distribution.
- Can be automated for ease of use.

Disadvantages:
- High initial investment and maintenance costs.
- Not suitable for irregularly shaped fields.

Best Practices:
- Monitor and adjust the system regularly for optimal coverage.
- Use low-pressure sprinklers to reduce energy use and evaporation.

Drip or Micro-Irrigation

1. Drip Irrigation

Description: Water is delivered directly to the root zone of plants through a network of tubes, emitters, and drippers.

Applications: Suitable for vegetable gardens, orchards, vineyards, and landscape plants.

Advantages:
- Highly efficient with minimal water loss.
- Reduces weed growth by targeting specific areas.

Disadvantages:
- High initial setup cost.
- Requires regular maintenance to prevent clogging.

Best Practices:
- Use pressure-compensating emitters for consistent water delivery.
- Install a filter to prevent clogging and ensure system longevity.

2. Micro-Sprinklers

Description: Small sprinklers deliver water in a fine mist or small droplets.

Applications: Suitable for orchards, nurseries, and flower beds.

Advantages:
- Provides gentle, uniform water distribution.
- Reduces soil erosion and surface runoff.

Disadvantages:
- Can be prone to clogging.
- Higher evaporation rates compared to drip irrigation.

Best Practices:
- Position micro-sprinklers to avoid wetting foliage and prevent disease.
- Use filtration systems to maintain clean water flow.

3. Soaker Hoses

Description: Porous hoses release water slowly along their length.

Applications: Suitable for flower beds, vegetable gardens, and hedges.

Advantages:
- Simple and inexpensive.
- Efficient water delivery directly to the soil.

Disadvantages:
- Limited to short distances.
- Requires regular maintenance to prevent clogging.

Best Practices:
- Bury soaker hoses under mulch to reduce evaporation.
- Regularly inspect and flush the hoses to maintain water flow.

Design Principles for Irrigation Systems

1. Site Assessment:
- Evaluate soil type, topography, and climate to select appropriate irrigation methods.
- Assess water sources and availability to ensure a reliable supply.

2. Water Efficiency:

- Design systems that minimize water loss through evaporation, runoff, and deep percolation.
- Use timers and sensors to optimize irrigation schedules based on plant needs and weather conditions.

3. Uniformity:

- Ensure even water distribution across the entire irrigated area.
- Use pressure regulators and flow control devices to maintain consistent water delivery.

4. Flexibility:

- Design systems that can be easily adjusted or expanded to accommodate changing plant needs or garden layout.
- Incorporate modular components for easy maintenance and upgrades.

5. Sustainability:
 - Integrate rainwater harvesting and greywater systems to supplement irrigation water.
 - Use renewable energy sources like solar pumps to power irrigation systems.

Best Practices for Irrigation Management

1. Regular Maintenance:
 - Inspect and maintain all components of the irrigation system regularly.
 - Clean filters, check for leaks, and replace worn or damaged parts.

2. Efficient Water Use:
 - Water plants early in the morning or late in the evening to reduce evaporation.
 - Adjust irrigation schedules based on weather conditions and soil moisture levels.

3. Soil Health:

- Improve soil structure and water-holding capacity with organic matter and mulch.
- Use cover crops to enhance soil fertility and reduce water runoff.

4. Monitoring and Adjustment:

- Use soil moisture sensors and weather data to optimize irrigation schedules.
- Monitor plant health and adjust watering as needed to prevent over- or under-watering.

5. Integrated Systems:

- Combine irrigation with other water management practices like mulching, composting, and crop rotation.
- Use companion planting and polycultures to enhance water efficiency and plant health.

Case Studies and Examples

Urban Garden Drip Irrigation:
- **Location:** A city rooftop garden.
- **System:** Drip irrigation with automated timers and rainwater harvesting.
- **Impact:** Reduced water use, healthier plants, and increased productivity.

Rural Farm Center Pivot Irrigation:
- **Location:** A large-scale agricultural farm.
- **System:** Center pivot irrigation with low-pressure sprinklers.
- **Impact:** Efficient water use, uniform crop growth, and increased yields.

Community Garden Soaker Hose System:
- **Location:** A suburban community garden.
- **System:** Soaker hoses with rainwater harvesting and greywater recycling.
- **Impact:** Improved water conservation, enhanced soil health, and community engagement.

By understanding and implementing appropriate irrigation techniques, permaculture gardeners can create sustainable, efficient, and productive systems that conserve water and promote healthy plant growth. Each irrigation method has its own benefits and challenges, and selecting the right one depends on factors like crop type, soil conditions, and available resources. With careful planning, regular maintenance, and integration with other sustainable practices, irrigation can play a key role in achieving a thriving permaculture garden.

Drip Irrigation

Drip irrigation is a highly efficient method of delivering water directly to the root zone of plants, reducing water wastage and promoting healthier plant growth. This guide explores the various aspects of drip irrigation, including its benefits, components, design principles, installation, maintenance, and best practices.

Introduction to Drip Irrigation

Drip irrigation, also known as micro-irrigation or trickle irrigation, involves the slow, precise application of water to the soil through a network of tubes, pipes, valves, and emitters. This method minimizes water loss due to evaporation, runoff, and deep percolation, making it an ideal choice for water-scarce regions and sustainable gardening practices.

Benefits of Drip Irrigation

1. Water Conservation: Drip irrigation uses 30-50% less water compared to traditional irrigation methods.

2. Improved Plant Health: By delivering water directly to the root zone, plants receive a steady supply of moisture, reducing stress and promoting healthier growth.

3. Reduced Weed Growth: Targeted watering minimizes moisture in non-planted areas, inhibiting weed germination.

4. Enhanced Fertilizer Efficiency: Nutrients can be applied directly to the root zone, reducing wastage and improving plant uptake.

5. Lower Disease Risk: Water is applied directly to the soil, keeping foliage dry and reducing the incidence of fungal diseases.

6. Scalability: Drip irrigation systems can be easily scaled to fit various garden sizes, from small vegetable patches to large agricultural fields.

Components of Drip Irrigation Systems

1. Water Source: The origin of the water supply, which can be a tap, rainwater tank, or pond.

2. Filter: Removes particles and debris to prevent clogging of emitters.

3. Pressure Regulator: Ensures consistent water pressure throughout the system to prevent damage and maintain uniform distribution.

4. Main Line: The primary conduit that delivers water from the source to the distribution lines.

5. Distribution Lines: Smaller tubes that branch off from the main line and carry water to the emitters.

6. Emitters/Drippers: Devices that release water at a controlled rate directly to the plant's root zone.

7. Fittings and Connectors: Various parts used to join, branch, and extend tubing, including tees, elbows, and couplings.

8. End Caps: Used to close off the ends of the distribution lines to ensure water flows through the emitters.

9. Timers: Automate watering schedules to ensure consistent irrigation.

Types of Drip Emitters

1. Inline Drippers: Integrated into the tubing and spaced at regular intervals.

2. End-Of-Line Drippers: Attached at the end of the tubing, allowing placement flexibility.

3. Pressure-Compensating Emitters: Deliver a consistent flow rate regardless of pressure variations.

4. Adjustable Emitters: Allow manual adjustment of the water flow rate to meet the needs of different plants.

5. Micro-Sprinklers and Sprayers: Provide a broader coverage area for plants with larger root zones.

Design Principles for Drip Irrigation Systems

1. Water Requirement Assessment:
   - Calculate the water needs of the plants based on their type, growth stage, and local climate.
   - Ensure the system is capable of meeting the peak water demand during the hottest periods.

2. Zoning:
- Divide the garden into zones based on plant type, water needs, and sun exposure.
- Each zone should have its own irrigation schedule to optimize water use.

3. Uniform Distribution:
- Design the system to ensure even water distribution across the entire irrigated area.
- Use pressure-compensating emitters to maintain consistent flow rates.

4. System Layout:
- Plan the layout to minimize the length of tubing and reduce pressure loss.
- Position emitters close to the plant's root zone for maximum efficiency.

5. Scalability and Flexibility:
- Design the system to allow for future expansion or modification.
- Use modular components for easy maintenance and upgrades.

Installation of Drip Irrigation Systems

1. Planning and Preparation:
 - Sketch a detailed layout of the garden, noting the location of plants, water sources, and potential obstacles.
 - Gather all necessary components, ensuring compatibility and adequate quantities.

2. Assembly of Main Line:
 - Install the main line from the water source to the garden, burying it slightly to protect it from damage.
 - Connect a filter and pressure regulator to the main line to ensure clean, consistent water flow.

3. Laying Distribution Lines:
 - Lay out the distribution lines along plant rows or beds, securing them with stakes or clips.

- Use fittings and connectors to branch off the main line and navigate around obstacles.

4. Installing Emitters:
- Attach emitters at appropriate intervals along the distribution lines, positioning them close to the plant's root zone.
- Ensure emitters are secure and not obstructed by soil or mulch.

5. Testing the System:
- Turn on the water supply and check for leaks, blockages, or uneven distribution.
- Adjust emitters and pressure settings as needed to ensure optimal performance.

6. Programming Timers:
- Set timers to automate the irrigation schedule, adjusting for seasonal variations and plant needs.
- Monitor the system regularly to ensure it is operating correctly and adjust the schedule as necessary.

Maintenance of Drip Irrigation Systems

1. Regular Inspection:
- Check the system weekly for leaks, blockages, and damaged components.
- Clean or replace clogged emitters to maintain efficient water delivery.

2. Seasonal Adjustments:
- Adjust the irrigation schedule based on seasonal changes in weather and plant growth stages.
- Winterize the system by draining water and protecting components from freezing temperatures.

3. System Flushing:
- Periodically flush the system to remove sediment and debris from the lines.
- Open end caps and run water through the system until it flows clear.

4. Filter Maintenance:

- Clean or replace filters regularly to prevent clogging and ensure clean water flow.
- Check filters after heavy rain or water source changes.

5. System Upgrades:

- Upgrade components as needed to improve efficiency and performance.
- Expand the system to accommodate new plantings or changes in garden layout.

Best Practices for Drip Irrigation

1. Water Quality:

- Use clean, filtered water to prevent clogging and damage to emitters.
- Install a filter at the water source to remove debris and particles.

2. Soil Health:

- Maintain healthy soil with organic matter and mulch to improve water retention and root development.
- Monitor soil moisture levels to avoid over- or under-watering.

3. Efficient Scheduling:

- Water early in the morning or late in the evening to reduce evaporation.
- Adjust watering frequency based on weather conditions and soil moisture levels.

4. Mulching:

- Use mulch to cover the soil and retain moisture, reducing evaporation and weed growth.
- Ensure mulch does not obstruct emitters or interfere with water distribution.

5. Plant Spacing:

- Space plants appropriately to ensure each receives adequate water from the emitters.

- Group plants with similar water needs together to optimize irrigation efficiency.

Case Studies and Examples

Urban Rooftop Garden:
- **Location:** A city rooftop garden with raised beds and container plants.
- **System:** Drip irrigation with pressure-compensating emitters and automated timers.
- **Impact:** Efficient water use, reduced labor, and healthier plants with increased yields.

Suburban Backyard Garden:
- **Location:** A suburban backyard with a mix of vegetable beds, fruit trees, and ornamental plants.
- **System:** Drip irrigation with inline drippers and micro-sprinklers for larger plants.
- **Impact:** Enhanced plant health, reduced water bills, and improved garden productivity.

Large-Scale Vineyard:
- **Location:** A vineyard in a semi-arid region.
- **System:** Drip irrigation with pressure-compensating emitters and moisture sensors.
- **Impact:** Optimized water use, improved grape quality, and increased resilience to drought conditions.

Conclusion

Drip irrigation is a highly effective and sustainable method of watering plants, offering numerous benefits in terms of water conservation, plant health, and garden productivity. By understanding the components, design principles, installation procedures, and maintenance practices, gardeners and farmers can implement drip irrigation systems that meet their specific needs and contribute to a more sustainable and efficient gardening approach. With careful planning and regular upkeep, drip irrigation can significantly enhance the success and sustainability of any garden or agricultural operation.

Swales and Keyline Design

Swales and Keyline Design

Swales and keyline design are integral elements of permaculture and sustainable land management practices. Both techniques focus on optimizing water distribution and retention in the landscape, promoting soil health, and enhancing ecosystem resilience. This guide explores the concepts, principles, benefits, and implementation of swales and keyline design, providing a detailed understanding of their roles in sustainable agriculture and land restoration.

Swales

Definition:
Swales are shallow, broad, and vegetated channels designed to capture, store, and infiltrate rainwater. Unlike traditional drainage ditches, swales slow down water flow, allowing it to soak into the soil rather than being quickly diverted away.

Principles of Swales:

1. Contour Alignment:
   - Swales are typically constructed along the contour lines of a landscape, meaning they follow the natural elevation contours of

the land. This helps to slow down and spread water evenly across the landscape.

2. Infiltration and Water Storage:

- The primary function of a swale is to capture and hold water, allowing it to infiltrate into the soil. This recharges groundwater, reduces surface runoff, and minimizes erosion.

3. Vegetation and Planting:

- Swales are often planted with perennial vegetation, including grasses, shrubs, and trees. The roots of these plants help stabilize the soil, increase water infiltration, and improve soil fertility.

Benefits of Swales:

1. Water Conservation:

- By capturing and storing rainwater, swales reduce the need for supplemental irrigation and enhance water availability for plants.

2. Erosion Control:
- Swales prevent soil erosion by slowing down water flow and reducing surface runoff.

3. Soil Improvement:
- The infiltration of water enhances soil moisture levels, promoting the growth of beneficial soil organisms and improving soil structure.

4. Habitat Creation:
- Vegetated swales provide habitat for wildlife, including insects, birds, and small mammals.

5. Flood Mitigation:
- Swales help mitigate flooding by reducing the volume and speed of runoff during heavy rain events.

Design and Construction of Swales:

1. Site Assessment:
- Evaluate the landscape to identify suitable locations for swales. Consider factors such as slope, soil type, and existing vegetation.

2. Contour Mapping:
- Use tools like an A-frame level, laser level, or contour maps to identify the contour lines of the landscape.

3. Excavation:
- Dig shallow, broad channels along the contour lines. The depth and width of the swale depend on the site conditions and the amount of water to be captured.

4. Berm Construction:
- Place the excavated soil on the downhill side of the swale to create a berm. The berm helps to hold water in the swale and can be planted with vegetation.

5. Vegetation:

- Plant the swale and berm with a mix of perennial grasses, shrubs, and trees. Select species that are well-adapted to the local climate and soil conditions.

6. Maintenance:

- Regularly inspect and maintain swales to ensure they function effectively. Remove any debris, repair any erosion, and manage vegetation as needed.

Keyline Design

Definition:

Keyline design is a holistic approach to land management developed by Australian farmer and engineer P.A. Yeomans. It focuses on optimizing water distribution and enhancing soil fertility through strategic plowing, water harvesting, and landscape planning.

Principles of Keyline Design:

1. Keyline and Keypoint:
- The keyline is a contour line that runs through the keypoint, which is the point of the most rapid change in slope in a valley. The keyline serves as the reference for designing the entire system.

2. Contour Plowing:
- Plowing along contour lines, especially above and below the keyline, helps to capture and direct water across the landscape, promoting even water distribution and infiltration.

3. Water Harvesting:
- Keyline design incorporates water harvesting techniques, such as dams, ponds, and channels, to capture and store rainwater for use during dry periods.

4. Soil Building:
- The strategic plowing and water management techniques in keyline design improve soil structure, increase organic matter, and enhance soil fertility.

5. Landscape Planning:
- Keyline design involves careful planning of land use, including the placement of roads, fences, and agricultural activities, to maximize efficiency and sustainability.

Benefits of Keyline Design:

1. Water Management:
- Optimizes water distribution and retention, reducing the need for supplemental irrigation and enhancing drought resilience.

2. Soil Fertility:
- Improves soil structure and organic matter content, promoting healthy plant growth and increased productivity.

3. Erosion Control:
- Reduces soil erosion by slowing down water flow and promoting infiltration.

4. Increased Productivity:
- Enhances the overall productivity of the land by improving soil health and water availability.

5. Sustainability:
- Promotes sustainable land management practices that enhance ecosystem resilience and biodiversity.

Design and Implementation of Keyline Design:

1. Site Assessment:
- Conduct a thorough assessment of the landscape, including topography, soil types, vegetation, and existing water features.

2. Keypoint Identification:

- Identify the key points in the landscape using contour maps or on-site observations. These are the points of the most rapid change in slope in valleys.

3. Keyline Mapping:

- Draw contour lines through the key points to establish the keylines. These lines serve as the basis for the design of the entire system.

4. Plowing and Cultivation:

- Plow along the contour lines above and below the keylines to capture and direct water across the landscape. Use specialized equipment like the Yeomans Plow for deep, non-inversion tillage.

5. Water Harvesting Structures:

- Design and construct water harvesting structures such as dams, ponds, and channels to capture and store rainwater.

6. Landscape Planning:

- Plan the placement of roads, fences, and agricultural activities based on the keyline design to maximize efficiency and sustainability.

7. Vegetation and Soil Improvement:

- Plants cover crops, trees, and other vegetation to improve soil health and promote water infiltration. Apply organic amendments to enhance soil fertility.

Integration of Swales and Keyline Design

Swales and keyline design can be integrated to create a comprehensive water management system that maximizes water retention, improves soil health, and enhances landscape resilience. Here's how these techniques can be combined effectively:

1. Complementary Placement:
   - Use swales to capture and infiltrate water along the contour lines, while keyline plowing directs excess water towards swales or water harvesting structures.

2. Enhanced Water Distribution:
   - Combine swales with keyline design to optimize water distribution across the landscape, ensuring even moisture levels and reducing runoff.

3. Synergistic Soil Improvement:
   - Integrate the soil-building benefits of keyline plowing with the organic matter contribution from swale vegetation to enhance soil health and fertility.

4. Biodiversity and Habitat Creation:
   - Use swales to create diverse planting zones and habitats for wildlife, while keyline design ensures sustainable land use and ecosystem resilience.

5. Holistic Land Management:
   - Implement both techniques as part of a holistic land management plan that considers water, soil, vegetation, and landscape planning to achieve long-term sustainability.

Case Studies and Examples

Swales in a Permaculture Garden:
- **Location**: A small permaculture garden in a suburban setting.
- **System:** A series of swales along contour lines with perennial vegetation and mulch.
- **Impact:** Improved water retention, reduced irrigation needs, and healthier plants with increased yields.

Keyline Design on a Farm:
- **Location:** A large-scale agricultural farm in a semi-arid region.
- **System**: Keyline plowing, water harvesting dams, and strategically placed vegetation.

- **Impact:** Enhanced soil fertility, increased water availability, and improved crop productivity.

Integrated Swales and Keyline Design in a Homestead:
- **Location:** A rural homestead with diverse agricultural activities.
- **System:** Swales for water infiltration and erosion control, combined with keyline plowing for optimal water distribution and soil improvement.
- **Impact:** Sustainable water management, increased biodiversity, and resilient land use practices.

Conclusion

Swales and keyline design are powerful tools for sustainable land management, offering numerous benefits in terms of water conservation, soil health, and ecosystem resilience. By understanding the principles, benefits, and implementation techniques of these methods, land managers, gardeners, and farmers can create landscapes that are both productive and sustainable. Integrating swales and keyline design into a comprehensive land management plan can lead to long-term improvements in soil fertility, water availability, and overall landscape health, contributing to a more sustainable and resilient future.

CHAPTER 4

Plant Selection

Plant Selection in Permaculture Gardening

Plant selection is a critical component of permaculture gardening. Choosing the right plants can significantly impact the health, productivity, and sustainability of your garden. This guide explores the principles of plant selection, types of plants suitable for permaculture, the importance of biodiversity, and specific strategies to optimize plant choices for various permaculture systems.

Principles of Plant Selection

1. Functional Diversity:
  - Select plants that serve multiple functions, such as providing food, enhancing soil fertility, attracting beneficial insects, and offering shade or shelter.

2. Adaptation to Local Conditions:
  - Choose plants that are well-suited to the local climate, soil type, and microclimate conditions. This increases the chances of successful growth and reduces the need for inputs like water, fertilizers, and pest control.

3. Succession Planning:
  - Plan for plant succession, which involves selecting plants that will occupy different ecological niches over time. This

ensures continuous productivity and resilience in the garden.

4. Companion Planting:
   - Incorporate companion planting principles, where plants are selected based on their beneficial interactions with each other. This can improve growth, reduce pest pressure, and enhance soil health.

5. Perennial Preference:
   - Emphasize perennial plants, which have longer life cycles and require less maintenance compared to annuals. Perennials contribute to soil stability, carbon sequestration, and long-term productivity.

6. Edibility and Utility:
   - Prioritize plants that provide edible yields or have other practical uses, such as medicinal herbs, fiber plants, or those that can be used for crafting and building materials.

7. Pollinator Support:
   - Include a variety of flowering plants that attract and support pollinators like bees, butterflies, and birds. This enhances pollination and overall garden health.

Types of Plants Suitable for Permaculture

1. Trees:
   - **Fruit Trees:** Apples, pears, plums, cherries, and other fruit-bearing trees provide food and habitat.
   - **Nut Trees:** Walnut, chestnut, and almond trees offer protein-rich nuts and improve soil through leaf litter.
   - **Nitrogen-Fixing Trees:** Species like alder, black locust, and acacia enrich the soil by fixing atmospheric nitrogen.

2. Shrubs:
   - **Berry Bushes:** Blueberries, raspberries, blackberries, and currants are

excellent for food production and attracting wildlife.

- **Medicinal Shrubs:** Elderberry, witch hazel, and rosemary have health benefits and can be used for natural remedies.

3. Herbs:

- **Culinary Herbs:** Basil, oregano, thyme, and parsley add flavor to food and have companion planting benefits.

- **Medicinal Herbs:** Echinacea, chamomile, and lavender provide health benefits and attract beneficial insects.

4. Vegetables:

- **Leafy Greens**: Kale, spinach, and Swiss chard are nutritious and easy to grow.

- **Root Vegetables:** Carrots, beets, and radishes help break up soil and provide edible roots.

- **Legumes:** Beans, peas, and lentils fix nitrogen and improve soil fertility.

5. Cover Crops and Green Manures:
   - **Cover Crops**: Clover, vetch, and rye protect soil from erosion and add organic matter.
   - **Green Manures:** Buckwheat, mustard, and lupine improve soil fertility when turned into the soil.

6. Flowers:
   - **Pollinator Plants:** Sunflowers, zinnias, and cosmos attract pollinators and beneficial insects.
   - **Medicinal Flowers:** Calendula, borage, and yarrow have medicinal properties and improve garden aesthetics.

Importance of Biodiversity

Biodiversity is a cornerstone of permaculture gardening. A diverse garden ecosystem is more resilient to pests, diseases, and environmental stress. Here are key aspects of biodiversity in plant selection:

1. Genetic Diversity:
   - Plant a variety of species and cultivars to increase genetic diversity. This enhances resilience and adaptability to changing conditions.

2. Functional Diversity:
   - Include plants with different roles, such as nitrogen fixers, nutrient accumulators, ground covers, and pollinator attractors. This creates a balanced and self-sustaining ecosystem.

3. Habitat Diversity:
   - Create diverse habitats within the garden, including hedgerows, ponds, and meadows. This supports a wide range of wildlife and beneficial organisms.

4. Temporal Diversity:
   - Ensure continuous production by selecting plants with different growing

seasons and life cycles. This provides year-round yields and habitat.

Strategies for Plant Selection

1. Climate and Microclimate Considerations:
   - Assess the microclimate of your region and the microclimates within your garden. Select plants that thrive in these specific conditions.

2. Soil Analysis:
   - Test your soil to understand its pH, texture, and nutrient content. Choose plants that are well-suited to your soil conditions or amend the soil to meet plant needs.

3. Zoning:
   - Use permaculture zoning principles to place plants according to their needs and your interaction frequency. Zone 1 might include herbs and vegetables, while zone 4 could have less frequently accessed trees.

4. Guilds:
- Design plant guilds, which are groups of plants that support each other's growth. A classic example is the "Three Sisters" guild of corn, beans, and squash.

5. Succession Planting:
- Plan for succession by including pioneer species that prepare the soil for longer-term plants. For instance, legumes can precede fruit trees to improve soil nitrogen levels.

6. Local and Native Plants:
- Prioritize native plants and those that are well-adapted to local conditions. These plants are more likely to thrive with minimal inputs and support local ecosystems.

7. Resource Availability:
- Consider the availability of water, sunlight, and other resources. Match plant selection to the resources available in different garden areas.

Examples of Plant Selection for Different Permaculture Systems

1. Forest Garden:

- **Canopy Layer:** Tall fruit and nut trees like apple, chestnut, and walnut.

- **Understory Layer:** Smaller fruit trees and shrubs like hazelnut, cherry, and berry bushes.

- **Herbaceous Layer:** Perennial herbs like comfrey, mint, and lemon balm.

- **Ground Cover Layer:** Clover, strawberries, and creeping thyme.

- **Root Layer:** Edible roots like Jerusalem artichoke and dandelion.

- **Climbing Layer:** Vining plants like grapes, kiwi, and beans.

2. Market Garden:

- **Annual Vegetables**: A mix of fast-growing crops like lettuce, radishes, and tomatoes.

- **Perennial Vegetables:** Asparagus, rhubarb, and perennial onions.
- **Herbs:** Both culinary and medicinal herbs like basil, rosemary, and calendula.
- **Cover Crops:** Clover and vetch to improve soil health and fertility between planting seasons.

3. Urban Permaculture Garden:
- **Vertical Gardening:** Utilize trellises and walls for vining plants like beans and cucumbers.
- **Container Gardening**: Grow herbs, leafy greens, and small root vegetables in pots.
- **Small Fruit Trees:** Dwarf varieties of apples, lemons, and figs.
- **Pollinator Plants:** Native flowering plants attract bees and butterflies.

4. Homestead Garden:
- **Diverse Food Production:** A mix of fruit trees, nut trees, vegetables, and herbs.

- **Animal Integration:** Plants that provide forage for chickens, ducks, and other livestock.

- **Medicinal Plants:** A variety of medicinal herbs for home remedies.

- **Water Management Plants:** Species that thrive in wet areas or help manage water flow, such as willows and rushes.

Conclusion

Plant selection in permaculture gardening is a nuanced and vital process that can determine the success and sustainability of the garden. By focusing on principles like functional diversity, local adaptation, and companion planting, gardeners can create resilient ecosystems that thrive with minimal inputs. The emphasis on biodiversity ensures a healthy, balanced environment that supports plants, wildlife, and beneficial organisms. Whether designing a forest garden, market garden, urban garden, or homestead, thoughtful plant selection lays the foundation for a productive and sustainable permaculture system.

1. Choosing the Right Plants

Choosing the Right Plants for Permaculture Gardening

Selecting the appropriate plants for a permaculture garden is crucial to creating a productive, sustainable, and resilient ecosystem. This comprehensive guide delves into the factors influencing plant selection, the types of plants suitable for various permaculture systems, and strategies to optimize plant choices for maximum benefit.

Factors Influencing Plant Selection

1. Climate and Microclimate:
 - Understanding the climate of your region, including temperature ranges,

precipitation patterns, and seasonal variations, is fundamental. Microclimates within your garden, such as shaded areas, sun traps, and frost pockets, also significantly influence plant choices.

2. Soil Characteristics:
   - Soil pH, texture, structure, and fertility are key factors. Conduct a soil test to determine these properties and select plants that thrive in your soil conditions. Amend the soil as necessary to meet the needs of particular plants.

3. Water Availability:
   - Assess the water availability and drainage in your garden. Choose drought-tolerant plants for dry areas and moisture-loving plants for wetter zones. Implement water management strategies like swales, rainwater harvesting, and mulching to optimize water use.

4. Sunlight:

- Different plants require varying amounts of sunlight. Map the sunlight exposure in your garden throughout the day and seasons to identify full-sun, partial-sun, and shaded areas. Select plants accordingly to ensure they receive the optimal amount of light.

5. Purpose and Function:

- Determine the primary functions you want your plants to serve. This includes providing food, improving soil fertility, attracting pollinators, offering shade, and more. Choose plants that fulfill multiple roles to maximize their utility in your garden.

6. Biodiversity:

- Promoting biodiversity is a cornerstone of permaculture. Select a diverse array of plants to create a resilient ecosystem that supports various beneficial organisms and mitigates pest and disease outbreaks.

7. Native and Local Adaptation:
- Prioritize native plants and those adapted to local conditions. These plants are typically more resilient, require fewer inputs, and support local wildlife and ecosystems.

8. Plant Life Cycles:
- Consider the life cycles of plants—annuals, biennials, and perennials. Incorporate a mix to ensure year-round productivity and ecological balance.

Types of Plants for Permaculture

1. Trees:
- **Fruit Trees**: Apples, pears, plums, and cherries are common choices. These trees provide food and habitat while contributing to soil health through leaf litter.
- **Nut Trees:** Walnuts, chestnuts, and almonds offer high-protein nuts and improve soil through organic matter.

- **Nitrogen-Fixing Trees:** Species like black locust, acacia, and alder enrich the soil by fixing atmospheric nitrogen.

2. Shrubs:
- **Berry Bushes:** Blueberries, raspberries, blackberries, and currants are excellent for food production and attracting wildlife.
- **Medicinal Shrubs:** Elderberry, witch hazel, and rosemary have health benefits and can be used for natural remedies.

3. Herbs:
- **Culinary Herbs**: Basil, oregano, thyme, and parsley add flavor to food and have companion planting benefits.
- **Medicinal Herbs**: Echinacea, chamomile, and lavender provide health benefits and attract beneficial insects.

4. Vegetables:
- **Leafy Greens:** Kale, spinach, and Swiss chard are nutritious and easy to grow.

- **Root Vegetables:** Carrots, beets, and radishes help break up soil and provide edible roots.
- **Legumes:** Beans, peas, and lentils fix nitrogen and improve soil fertility.

5. Cover Crops and Green Manures:
- **Cover Crops**: Clover, vetch, and rye protect soil from erosion and add organic matter.
- **Green Manures**: Buckwheat, mustard, and lupine improve soil fertility when turned into the soil.

6. Flowers:
- **Pollinator Plants:** Sunflowers, zinnias, and cosmos attract pollinators and beneficial insects.
- **Medicinal Flowers:** Calendula, borage, and yarrow have medicinal properties and improve garden aesthetics.

Strategies for Plant Selection

1. Climate and Microclimate Considerations:
   - Assess the macroclimate of your region and the microclimates within your garden. Select plants that thrive in these specific conditions.

2. Soil Analysis:
   - Test your soil to understand its pH, texture, and nutrient content. Choose plants that are well-suited to your soil conditions or amend the soil to meet plant needs.

3. Zoning:
   - Use permaculture zoning principles to place plants according to their needs and your interaction frequency. Zone 1 might include herbs and vegetables, while zone 4 could have less frequently accessed trees.

4. Guilds:

- Design plant guilds, which are groups of plants that support each other's growth. A classic example is the "Three Sisters" guild of corn, beans, and squash.

5. Succession Planting:

- Plan for succession by including pioneer species that prepare the soil for longer-term plants. For instance, legumes can precede fruit trees to improve soil nitrogen levels.

6. Local and Native Plants:

- Prioritize native plants and those that are well-adapted to local conditions. These plants are more likely to thrive with minimal inputs and support local ecosystems.

7. Resource Availability:

- Consider the availability of water, sunlight, and other resources. Match plant selection to the resources available in different garden areas.

Examples of Plant Selection for Different Permaculture Systems

1. Forest Garden:

- **Canopy Layer**: Tall fruit and nut trees like apple, chestnut, and walnut.
- **Understory Layer:** Smaller fruit trees and shrubs like hazelnut, cherry, and berry bushes.
- **Herbaceous Layer:** Perennial herbs like comfrey, mint, and lemon balm.
- **Ground Cover Layer**: Clover, strawberries, and creeping thyme.
- **Root Layer**: Edible roots like Jerusalem artichoke and dandelion.
- **Climbing Layer:** Vining plants like grapes, kiwi, and beans.

2. Market Garden:

- **Annual Vegetables**: A mix of fast-growing crops like lettuce, radishes, and tomatoes.
- **Perennial Vegetables:** Asparagus, rhubarb, and perennial onions.

- **Herbs:** Both culinary and medicinal herbs like basil, rosemary, and calendula.
- **Cover Crops:** Clover and vetch to improve soil health and fertility between planting seasons.

3. Urban Permaculture Garden:

- **Vertical Gardening**: Utilize trellises and walls for vining plants like beans and cucumbers.
- **Container Gardening:** Grow herbs, leafy greens, and small root vegetables in pots.
- **Small Fruit Trees:** Dwarf varieties of apples, lemons, and figs.
- **Pollinator Plants**: Native flowering plants to attract bees and butterflies.

4. Homestead Garden:

- **Diverse Food Production:** A mix of fruit trees, nut trees, vegetables, and herbs.
- **Animal Integration**: Plants that provide forage for chickens, ducks, and other livestock.

- **Medicinal Plants:** A variety of medicinal herbs for home remedies.
- **Water Management Plants**: Species that thrive in wet areas or help manage water flow, such as willows and rushes.

Specific Plant Considerations

1. Drought-Tolerant Plants:
- Select plants like lavender, rosemary, and sage for dry areas. These plants require minimal water and thrive in well-drained soils.

2. Shade-Tolerant Plants:
- For shaded areas, choose plants like hostas, ferns, and certain varieties of leafy greens that perform well with limited sunlight.

3. Windbreak and Shelter Plants:
- Use tall, hardy plants like bamboo, pine, and cypress to create windbreaks and protect more delicate plants.

4. Pioneer Species:
   - Incorporate fast-growing pioneer species like legumes and sunflowers that quickly establish and improve soil conditions for future plantings.

5. Nitrogen Fixers:
   - Include nitrogen-fixing plants such as clover, alfalfa, and legumes to enhance soil fertility and support the growth of surrounding plants.

6. Pollinator-Friendly Plants:
   - Plant a variety of flowers like marigolds, cosmos, and echinacea to attract pollinators and beneficial insects.

7. Medicinal and Culinary Herbs:
   - Grow a diverse range of herbs like thyme, basil, and mint for culinary uses, and others like echinacea and chamomile for their medicinal properties.

8. Edible Perennials:
   - Incorporate perennial vegetables like asparagus, rhubarb, and artichokes for long-term food production.

Conclusion

resilient permaculture garden that will continue to provide benefits for years to come. By fostering biodiversity, promoting ecological balance, and optimizing the natural resources available, you not only create a garden that meets your needs but also contributes positively to the environment. The right plant choices can transform your garden into a self-sustaining ecosystem, rich in productivity, and harmony with nature. This careful selection process ensures that your permaculture garden is not only a source of food and beauty but also a model of sustainable living and ecological stewardship.

Native and Adaptable Species

Native and Adaptable Species in Permaculture Gardening

Choosing the right plants for a permaculture garden involves more than just aesthetics or yield. It requires a deep understanding of the ecosystem and the selection of species that are both native and adaptable to local conditions. This approach ensures sustainability, resilience, and ecological harmony. This comprehensive guide explores the importance of native and adaptable species, their benefits, and how to effectively incorporate them into a permaculture garden.

Importance of Native Species

1. Ecological Compatibility:
   - Native plants are inherently suited to the local environment. They have evolved over

thousands of years in tandem with the local climate, soil, and wildlife, making them perfectly adapted to these conditions.

2. Support for Local Wildlife:

- Native plants provide essential resources such as nectar, pollen, seeds, and habitat for local wildlife, including birds, insects, and mammals. This helps maintain biodiversity and ecological balance.

3. Reduced Maintenance:

- Because they are adapted to the local environment, native plants generally require less water, fertilizers, and pesticides compared to non-native species. This makes them easier to care for and more sustainable.

4. Resilience to Local Pests and Diseases:

- Native plants have developed natural defenses against local pests and diseases, reducing the need for chemical interventions and promoting a healthier garden ecosystem.

Benefits of Adaptable Species

1. Climate Versatility:
   - Adaptable species can thrive in a range of environmental conditions. This flexibility makes them valuable for gardens experiencing varied weather patterns or those undergoing climate change.

2. Enhanced Biodiversity:
   - Introducing adaptable species can enhance biodiversity by filling ecological niches not covered by native plants. This increases the resilience and stability of the garden ecosystem.

3. Resource Efficiency:
   - Adaptable plants often make efficient use of available resources such as water, sunlight, and nutrients, making them

suitable for permaculture systems designed to maximize resource use efficiency.

4. Complementing Native Species:
- Adaptable species can complement native plants by providing additional benefits such as improved soil fertility, pest control, or extended harvest periods.

Strategies for Incorporating Native and Adaptable Species

1. Research and Planning:
- Begin with thorough research on the native flora of your region. Local botanical gardens, extension services, and native plant societies are excellent resources for this information.

2. Assessing Local Conditions:
- Conduct a detailed assessment of your garden's soil, climate, microclimates, and water availability. This helps in selecting

plants that will thrive in your specific conditions.

3. Creating Plant Guilds:
   - Design plant guilds that incorporate native and adaptable species working together to support each other's growth. For example, a fruit tree (native or adaptable) surrounded by nitrogen-fixing plants, ground covers, and pollinator-attracting flowers.

4. Phased Introduction:
   - Introduce new plants in phases to observe their adaptability and impact on the existing ecosystem. This allows for adjustments and ensures successful integration.

5. Local and Regional Focus:
   - Prioritize plants that are not only native to your country but specifically to your region. This ensures the highest level of adaptation and ecological benefit.

6. Community Involvement:

- Engage with local gardening communities, permaculture groups, and indigenous knowledge holders. They can provide insights and recommendations for suitable native and adaptable species.

Examples of Native and Adaptable Species for Various Regions

1. North America:
- **Native**: Black-eyed Susan (Rudbeckia hirta), Eastern Redbud (Cercis canadensis), Purple Coneflower (Echinacea purpurea).
- **Adaptable:** Sunflowers (Helianthus spp.), Swiss Chard (Beta vulgaris), Lavender (Lavandula spp.).

2. Europe:
- **Native:** Common Heather (Calluna vulgaris), European Beech (Fagus sylvatica), Yarrow (Achillea millefolium).

- **Adaptable**: Rosemary (Rosmarinus officinalis), Tomatoes (Solanum lycopersicum), Marigolds (Tagetes spp.).

3. Australia:

- **Native**: Kangaroo Paw (Anigozanthos spp.), Lemon Myrtle (Backhousia citriodora), Banksia (Banksia spp.).

- **Adaptable**: Basil (Ocimum basilicum), Zucchini (Cucurbita pepo), Nasturtium (Tropaeolum majus).

4. Africa:

- **Native:** Spekboom (Portulacaria afra), African Tulip Tree (Spathodea campanulata), Wild Dagga (Leonotis leonurus).

- **Adaptable**: Amaranth (Amaranthus spp.), Sweet Potatoes (Ipomoea batatas), Marigolds (Tagetes spp.).

5. Asia:
- **Native**: Japanese Maple (Acer palmatum), Lotus (Nelumbo nucifera), Bamboo (Bambusoideae).
- **Adaptable:** Eggplants (Solanum melongena), Peppers (Capsicum spp.), Cilantro (Coriandrum sativum).

6. South America:
- **Native**: Passionflower (Passiflora spp.), Yacon (Smallanthus sonchifolius), Jacaranda (Jacaranda mimosifolia).
- **Adaptable:** Corn (Zea mays), Beans (Phaseolus spp.), Sunflowers (Helianthus spp.).

Practical Steps for Planting and Maintenance

1. Soil Preparation:
- Ensure the soil is well-prepared with organic matter to support plant health. This includes composting, mulching, and

possibly adjusting pH levels to suit specific plants.

2. Planting Techniques:
   - Follow best practices for planting, such as proper spacing, planting depth, and timing. Native plants often have specific planting windows to align with natural growth cycles.

3. Watering Practices:
   - Establish a watering schedule that aligns with the needs of both native and adaptable species. Native plants often require less frequent watering once established.

4. Mulching:
   - Use organic mulch to retain soil moisture, regulate temperature, and suppress weeds. Mulching also provides a habitat for beneficial organisms.

5. Pest Management:
- Implement integrated pest management (IPM) strategies. Encourage beneficial insects, use companion planting, and minimize chemical interventions.

6. Monitoring and Adaptation:
- Regularly monitor plant health and growth. Be prepared to make adjustments based on observations, such as relocating plants or modifying care routines.

Benefits of Native and Adaptable Species

1. Sustainability:
- By choosing plants that are naturally suited to local conditions, you reduce the need for external inputs such as water, fertilizers, and pesticides. This enhances the sustainability of your garden.

2. Biodiversity:
- A diverse garden with a mix of native and adaptable species supports a wide range of organisms, from soil microbes to birds and pollinators, creating a vibrant and dynamic ecosystem.

3. Resilience:
- Gardens with a high diversity of native and adaptable plants are more resilient to pests, diseases, and environmental stresses like droughts or extreme weather events.

4. Aesthetic and Cultural Value:
- Native plants often have significant cultural and aesthetic value. They can connect gardeners to the natural heritage of their region and enhance the beauty of the garden.

5. Educational Opportunities:
- A garden rich in native and adaptable species offers educational opportunities to

learn about local ecosystems, plant biology, and sustainable gardening practices.

Conclusion

Incorporating native and adaptable species into a permaculture garden is a thoughtful and strategic approach that promotes sustainability, resilience, and ecological harmony. By understanding the specific conditions of your garden and selecting plants that are well-suited to these conditions, you can create a thriving, self-sustaining ecosystem. This approach not only benefits the environment but also enhances the productivity and beauty of your garden, providing lasting rewards for both the gardener and the broader ecosystem. Through careful planning, research, and ongoing observation, your permaculture garden can become a model of biodiversity and sustainable living.

Companion Planting

Companion Planting in Permaculture Gardening

Companion planting is a cornerstone of permaculture gardening, offering a natural and synergistic approach to plant cultivation. By strategically placing plants together, gardeners can create mutually beneficial relationships that enhance growth, deter pests, improve soil health, and increase yields. This comprehensive guide delves into the principles of companion planting, its benefits, common plant pairings, and practical tips for implementation.

Principles of Companion Planting

1. Mutual Benefit:
- The core principle of companion planting is mutual benefit. Plants are chosen to support each other's growth by providing nutrients, shade, support, or pest protection.

2. Biodiversity:
- Promoting biodiversity through companion planting creates a more resilient ecosystem. Diverse plantings attract beneficial insects, reduce the spread of pests and diseases, and enhance soil health.

3. Functional Relationships:

- Companion plants are selected based on their functional relationships, such as nutrient cycling, physical support, or microclimate modification. These relationships create a more efficient and productive garden.

4. Resource Optimization:

- Effective companion planting optimizes the use of resources such as water, sunlight, and soil nutrients, reducing competition and promoting overall plant health.

Benefits of Companion Planting

1. Pest Control:

- Certain plants repel pests or attract beneficial insects that prey on harmful pests. For example, marigolds are known to repel nematodes, while dill attracts predatory insects that control aphids.

2. Improved Soil Health:
- Some plants, such as legumes, fix nitrogen in the soil, improving fertility for neighboring plants. Others, like deep-rooted plants, bring up nutrients from deeper soil layers.

3. Enhanced Growth and Yield:
- Companion planting can enhance the growth and yield of crops. For instance, basil is known to improve the flavor and growth of tomatoes when planted nearby.

4. Efficient Space Utilization:
- By combining plants with different growth habits, such as tall plants with ground covers, gardeners can make better use of available space, maximizing productivity.

5. Microclimate Creation:
- Companion plants can modify the microclimate, providing shade, wind protection, or humidity control, which benefits sensitive plants.

Common Companion Plant Pairings

1. Three Sisters (Corn, Beans, and Squash):

- This traditional Native American trio involves planting corn, beans, and squash together. Corn provides support for climbing beans, beans fix nitrogen in the soil, and squash acts as a living mulch to suppress weeds and retain moisture.

2. Tomatoes and Basil:

- Basil improves the growth and flavor of tomatoes and repels pests like aphids and whiteflies. Both plants also benefit from similar growing conditions.

3. Carrots and Onions:

- Onions repel carrot flies, while carrots improve soil structure for onions. This pairing helps both crops thrive with reduced pest pressure.

4. Cabbage and Dill:
- Dill attracts beneficial insects that prey on cabbage pests like cabbage loopers and aphids. Dill also improves the flavor of cabbage.

5. Radishes and Cucumbers:
- Radishes act as a trap crop, attracting pests away from cucumbers. They also help to loosen the soil, benefiting cucumber roots.

6. Marigolds and Most Vegetables:
- Marigolds repel a variety of pests, including nematodes, aphids, and whiteflies. They are often planted alongside many vegetables to provide broad-spectrum pest control.

7. Strawberries and Borage:
- Borage attracts pollinators and improves the growth and flavor of strawberries. It also repels common strawberry pests.

Practical Tips for Implementing Companion Planting

1. Plan Your Garden Layout:
- Before planting, create a garden layout that pairs compatible plants. Consider each plant's mature size, growth habit, and resource needs.

2. Research Plant Interactions:
- Not all plants are good companions. Research specific plant interactions to avoid negative effects, such as allelopathy, where one plant inhibits the growth of another.

3. Rotate Crops:
- Implement crop rotation to prevent the buildup of pests and diseases. Rotate plant families to different garden areas each season.

4. Use Succession Planting:
- Practice succession planting to maintain continuous crop production and avoid pest

and disease cycles. Follow one crop with a complementary companion crop.

5. Incorporate Flowers and Herbs:
   - Include flowers and herbs that attract beneficial insects and provide pest control. These plants can be interspersed throughout the garden.

6. Observe and Adjust:
   - Regularly observe your garden to see how plants are interacting. Be prepared to make adjustments if certain pairings are not working well.

7. Maintain Diversity:
   - Maintain a diverse mix of plants to enhance resilience and reduce the risk of pest and disease outbreaks. Diversity also supports a healthy soil microbiome.

Companion Planting Examples and Their Benefits

1. Nitrogen Fixing:
   - Example: Peas and beans (legumes) with corn or spinach.
   - Benefit: Legumes fix atmospheric nitrogen, enriching the soil and benefiting neighboring plants.

2. Trap Cropping:
   - Example: Nasturtiums with brassicas (cabbage, broccoli).
   - Benefit: Nasturtiums attract aphids away from brassicas, reducing pest damage.

3. Physical Support:
   - Example: Sunflowers with climbing beans.
   - Benefit: Sunflowers provide a natural trellis for beans to climb, saving space and enhancing growth.

4. Beneficial Insect Attraction:

- Example: Yarrow with fruit trees.
- Benefit: Yarrow attracts beneficial insects that prey on pests, protecting the fruit trees.

5. Living Mulch:

- Example: Clover with corn or fruit trees.
- Benefit: Clover acts as a living mulch, suppressing weeds, retaining soil moisture, and fixing nitrogen.

6. Microclimate Modification:

- Example: Lettuce with taller plants like tomatoes.
- Benefit: Lettuce benefits from the shade provided by taller plants, reducing bolting in hot weather.

7. Pest Repellents:

- Example: Garlic with roses.
- Benefit: Garlic repels aphids and other pests that commonly affect roses.

Conclusion

Companion planting is a powerful technique in permaculture gardening that harnesses the natural relationships between plants to create a more sustainable, productive, and resilient garden. By understanding and implementing the principles of companion planting, gardeners can optimize plant health, reduce pest and disease pressure, improve soil fertility, and maximize yields. This method not only enhances the efficiency of the garden but also contributes to ecological balance and biodiversity. With thoughtful planning, research, and observation, companion planting can transform a garden into a thriving, self-sustaining ecosystem.

2. Perennial vs. Annual Plants

Perennial vs. Annual Plants in Permaculture Gardening

In permaculture gardening, the choice between perennial and annual plants plays a crucial role in shaping the garden's design, productivity, and sustainability. Both types of plants offer distinct advantages and challenges, and understanding their characteristics can help gardeners make informed decisions to create a balanced, resilient, and efficient garden system. This comprehensive guide explores the differences between perennial and annual plants, their benefits and drawbacks, and strategies for incorporating both types into a permaculture garden.

Understanding Perennial Plants

1. Definition and Characteristics:
   - Perennial plants are those that live for more than two years. They typically go through cycles of growth, dormancy, and regrowth, often producing flowers and seeds multiple times over their lifespan.

2. Growth and Maintenance:
   - Perennials have deep root systems that help them access water and nutrients from deeper soil layers. This characteristic often makes them more drought-tolerant and less dependent on frequent watering and fertilization.

3. Longevity:
- Perennials can live for many years, sometimes even decades. This longevity reduces the need for replanting and allows for the development of a more stable and permanent garden structure.

4. Examples:
- Common perennial plants include fruit trees (apple, pear), shrubs (blueberries, currants), herbs (sage, thyme), and vegetables (asparagus, rhubarb).

Benefits of Perennial Plants

1. Soil Health:
- The deep roots of perennials help improve soil structure, increase organic matter, and prevent erosion. They also contribute to soil fertility by bringing up nutrients from deeper layers.

2. Reduced Maintenance:

- Perennials require less frequent planting, weeding, and tilling, reducing labor and soil disturbance. This lower maintenance approach is well-suited to sustainable gardening practices.

3. Drought Tolerance:

- Deep root systems enable perennials to access water during dry periods, making them more resilient to drought conditions compared to shallow-rooted annuals.

4. Wildlife Support:

- Perennials provide consistent habitat and food sources for beneficial insects, birds, and other wildlife, promoting biodiversity and ecological balance.

5. Long-Term Productivity:

- Once established, perennials can provide reliable yields year after year, ensuring a steady supply of food and other resources.

Challenges of Perennial Plants

1. Initial Establishment:
   - Perennials often take longer to establish and reach full productivity compared to annuals. This requires patience and long-term planning.

2. Space Requirements:
   - Many perennials require more space and careful placement to avoid overcrowding and competition with other plants.

3. Management:
   - Perennials may need regular pruning, division, and care to maintain their health and productivity over time.

4. Pest and Disease Management:
   - Perennials can accumulate pests and diseases over the years, requiring vigilant monitoring and management practices.

Understanding Annual Plants

1. Definition and Characteristics:
 - Annual plants complete their life cycle—from germination to seed production—within one growing season. They die after producing seeds and must be replanted each year.

2. Growth and Maintenance:
 - Annuals typically have fast growth rates, allowing for quick establishment and harvest. They often have shallow root systems compared to perennials.

3. Short Lifespan:
 - The short lifespan of annuals means they need to be replanted every year, which involves more labor and soil disturbance.

4. Examples:
 - Common annual plants include vegetables (tomatoes, lettuce, beans),

flowers (marigolds, zinnias), and herbs (basil, cilantro).

Benefits of Annual Plants

1. Rapid Growth and Harvest:
   - Annuals grow quickly and can produce harvestable yields within a single growing season, making them ideal for immediate food production.

2. Flexibility and Crop Rotation:
   - The annual lifecycle allows for crop rotation, which helps manage soil fertility, reduce pest and disease buildup, and optimize space use.

3. Variety and Experimentation:
   - Gardeners can experiment with a wide variety of annual crops each year, adapting to changing tastes, climate conditions, and garden needs.

4. Soil Cover and Green Manure:

- Annual cover crops can improve soil fertility and structure, provide green manure, and suppress weeds during the off-season.

Challenges of Annual Plants

1. Labor-Intensive:

- Annuals require more frequent planting, weeding, and soil preparation, increasing labor and resource inputs.

2. Soil Disturbance:

- Regular planting and tilling can disrupt soil structure, reduce organic matter, and contribute to erosion.

3. Resource Dependence:

- Annuals often need more water, fertilizers, and pest control measures due to their fast growth and shallow roots.

4. Short-Term Focus:

- The focus on short-term productivity can sometimes lead to less consideration of long-term garden health and sustainability.

Integrating Perennial and Annual Plants in Permaculture

1. Design for Diversity:

- Incorporate a mix of perennials and annuals to create a diverse and resilient garden. Use perennials to provide long-term structure and stability, while annuals offer flexibility and quick yields.

2. Layering and Guilds:

- Design plant guilds that combine perennials and annuals. For example, plant annual vegetables under fruit trees or intersperse annual herbs with perennial shrubs.

3. Succession Planting:
- Use succession planting to maintain continuous production. Follow annual crops with perennials or rotate different annuals to optimize space and resources.

4. Soil Building:
- Use annual cover crops and green manures to build soil fertility and structure between perennial plantings. This helps maintain healthy, productive soil.

5. Pest and Disease Management:
- Diversify plantings to reduce pest and disease pressure. Perennials can provide habitat for beneficial insects, while annuals can be rotated to break pest and disease cycles.

6. Water Management:
- Combine the water efficiency of perennials with the flexibility of annuals. Use deep-rooted perennials to access subsoil

moisture and shallow-rooted annuals to take advantage of surface water.

7. Adaptation to Climate and Microclimates:

- Select perennials and annuals suited to your climate and microclimates. Perennials can provide shade and wind protection for annuals, creating favorable growing conditions.

Examples of Integrated Plantings

1. Perennial Herb Garden with Annuals:

- Combine perennial herbs like rosemary, thyme, and sage with annual herbs like basil, cilantro, and dill. This creates a year-round supply of herbs with minimal maintenance.

2. Fruit Tree Guilds:

- Design fruit tree guilds that include nitrogen-fixing plants (e.g., clover), pest-repelling herbs (e.g., garlic), and annual vegetables (e.g., squash) for a multifunctional, productive system.

3. Vegetable Beds with Perennial Borders:
   - Plant annual vegetables in beds surrounded by perennial borders of berries, asparagus, or rhubarb. The perennials provide habitat and support for the annuals.

4. Polyculture Plantings:
   - Create polycultures that combine multiple annual and perennial species. For example, interplant tomatoes, peppers, and eggplants (annuals) with chives, lavender, and oregano (perennials).

Conclusion

Understanding the differences between perennial and annual plants and their respective benefits and challenges is essential for creating a balanced and sustainable permaculture garden. By strategically integrating both types of plants, gardeners can maximize productivity, enhance soil health, and promote ecological balance. Perennials offer long-term stability, reduced maintenance, and environmental benefits, while annuals provide rapid yields, flexibility, and opportunities for crop rotation. Through thoughtful planning and design, combining perennials and annuals can lead to a resilient and efficient garden that supports both short-term and long-term goals. This integrated approach not only meets immediate needs but also contributes to the overall health and sustainability of the garden ecosystem.

Benefits of Perennials

Benefits of Perennials in Permaculture Gardening

Perennials play a pivotal role in permaculture gardening, offering a range of ecological, economic, and practical benefits. Their unique characteristics make them ideal for sustainable and resilient garden systems. This comprehensive guide explores the various advantages of incorporating perennials into your permaculture garden, from enhancing soil health and biodiversity to reducing maintenance and improving long-term productivity.

Ecological Benefits

1. Soil Health and Structure:
 - Deep Root Systems: Perennial plants develop extensive root systems that

penetrate deep into the soil. These roots help to improve soil structure by creating channels for air and water infiltration, enhancing soil porosity, and preventing compaction.

- **Organic Matter Contribution:** Perennials contribute organic matter to the soil through leaf litter, root exudates, and decomposition of plant parts. This organic matter increases soil fertility, supports beneficial microorganisms, and improves soil moisture retention.

- **Erosion Control:** The dense and permanent root systems of perennials stabilize the soil, reducing erosion caused by wind and water. This is particularly important on slopes and in areas prone to heavy rainfall.

2. Water Management:

- **Drought Tolerance:** Deep-rooted perennials are more capable of accessing water from deeper soil layers, making them more resilient during dry periods. This reduces the need for frequent watering and helps maintain plant health during droughts.

- **Water Conservation:** Perennials reduce water runoff and increase water infiltration, helping to recharge groundwater supplies and maintain a stable water table.

3. Biodiversity and Habitat:

- **Wildlife Support:** Perennial plants provide consistent habitat and food sources for a variety of wildlife, including pollinators (bees, butterflies), birds, and beneficial insects. This promotes biodiversity and ecological balance within the garden.

- **Pest Management**: By attracting beneficial insects and predators, perennials help control pest populations naturally. This reduces the need for chemical pesticides and supports a healthy garden ecosystem.

4. Carbon Sequestration:

- **Carbon Storage:** Perennials contribute to carbon sequestration by storing carbon in their biomass and root systems. This helps mitigate climate change by reducing atmospheric CO_2 levels.

Economic Benefits

1. Cost Efficiency:
- Reduced Planting Costs: Perennials do not need to be replanted every year, which reduces the cost of seeds, seedlings, and planting labor. Once established, they continue to produce for many years with minimal input.
- **Lower Maintenance Expenses:** The low-maintenance nature of perennials translates into savings on labor, water, fertilizers, and pest control measures.

2. Long-Term Productivity:
- **Consistent Yields:** Perennials provide reliable yields year after year, ensuring a steady supply of food, herbs, and other resources. This consistency is particularly valuable for home gardeners and small-scale farmers.

- **Market Value**: Perennial crops such as fruit trees, berry bushes, and nut trees often have higher market value compared to annual vegetables. This can provide a stable and profitable income stream.

Practical Benefits

1. Reduced Labor:
   - **Less Frequent Planting**: Perennials eliminate the need for annual planting, reducing labor associated with soil preparation, planting, and initial care. This is especially beneficial for gardeners with limited time or physical ability.
   - **Weed Suppression:** Established perennial plants can outcompete weeds, reducing the need for weeding and herbicide use. Ground-cover perennials, in particular, create a dense canopy that inhibits weed growth.

2. Season Extension:

- **Early and Late Harvests:** Some perennials, such as asparagus and rhubarb, can be harvested in early spring before annual crops are ready. Others, like certain fruit trees, extend the harvest season into late fall.
- **Year-Round Availability:** Many perennial herbs and greens are available year-round, providing fresh produce even during the off-season.

3. Resilience and Adaptability:

- **Climate Adaptation:** Perennials are often more adaptable to varying climate conditions due to their established root systems and long lifespans. They can better withstand temperature fluctuations, extreme weather, and changing growing conditions.
- **Reduced Risk:** The long-term nature of perennials spreads the risk of crop failure over several years. This reduces the impact of adverse weather, pest outbreaks, or other challenges that may affect annual crops.

Health and Nutritional Benefits

1. Nutrient Density:

- **High Nutritional Value:** Many perennial crops, such as berries, nuts, and leafy greens, are rich in vitamins, minerals, and antioxidants. Including a variety of perennials in the diet can enhance overall nutrition and health.

- **Soil Health Correlation:** The improved soil health associated with perennial plants often translates into more nutrient-dense produce. Healthy soil supports plants in producing higher quality and more nutritious food.

2. Accessibility:

- **Convenient Harvesting**: Perennials often produce over an extended period, allowing for continuous and convenient harvesting. This is especially useful for busy gardeners who prefer to harvest fresh produce as needed.

- **Food Security:** The reliability and ease of growing perennials contribute to food security by ensuring a steady and accessible supply of nutritious food.

Examples of Beneficial Perennial Plants

1. Fruit Trees:
 - **Examples**: Apple, pear, peach, cherry.
 - Benefits: Provide long-term yields of fresh fruit, support wildlife, and contribute to soil health through leaf litter.

2. Berry Bushes:
 - **Examples**: Blueberry, raspberry, blackberry, currant.
 - Benefits: Offer nutrient-dense berries, attract pollinators, and improve soil structure with deep root systems.

3. Nut Trees:
 - **Examples**: Walnut, hazelnut, chestnut.

- Benefits: Produce high-protein nuts, sequester carbon, and provide habitat for wildlife.

4. Perennial Vegetables:
- **Examples**: Asparagus, rhubarb, artichoke, perennial kale.
- Benefits: Yield early spring vegetables, improve soil fertility, and reduce annual planting needs.

5. Herbs:
- **Examples**: Rosemary, thyme, sage, oregano.
- Benefits: Offer culinary and medicinal uses, attract beneficial insects, and require minimal maintenance.

6. Ground Covers:
- Examples: Clover, creeping thyme, comfrey.
- Benefits: Suppress weeds, improve soil health, and provide habitat for beneficial insects.

Conclusion

Incorporating perennials into a permaculture garden offers a multitude of benefits, from enhancing soil health and biodiversity to reducing maintenance and improving long-term productivity. Their ecological, economic, practical, and health advantages make them an essential component of sustainable gardening practices. By understanding and leveraging the strengths of perennial plants, gardeners can create resilient, efficient, and productive garden systems that support both immediate and long-term goals. This integrated approach not only ensures a steady supply of nutritious food but also contributes to the overall health and sustainability of the garden ecosystem.

Integrating Annuals into Your Garden

Integrating Annuals into Your Permaculture Garden

Annual plants, despite their short life cycles, play a significant role in permaculture gardening. They offer quick yields, flexibility, and the ability to adapt to changing garden conditions. Integrating annuals into a permaculture garden requires strategic planning and thoughtful design to maximize their benefits while maintaining the overall sustainability and resilience of the system. This comprehensive guide explores the advantages of annuals, how to incorporate them effectively, and best practices for managing them in harmony with perennial plants.

Benefits of Annual Plants

1. Rapid Growth and Productivity:

- Quick Harvests: Annuals complete their life cycle within a single growing season, allowing for rapid production of vegetables, herbs, and flowers. This quick turnaround is ideal for gardeners seeking immediate results.

- Multiple Crops per Year: In many climates, it is possible to grow multiple crops of annuals in a single year, enhancing overall garden productivity.

2. Flexibility and Adaptability:

- Crop Rotation: Annuals facilitate crop rotation, which helps manage soil fertility, reduce pest and disease buildup, and optimize space use. Rotating crops prevents the depletion of specific nutrients and interrupts pest life cycles.

- Seasonal Variety: The wide variety of annual crops allows gardeners to adapt to changing tastes, climate conditions, and garden needs. This flexibility enables the cultivation of a diverse array of plants throughout the year.

3. Soil Improvement:

- **Cover Crops and Green Manures:** Certain annuals, such as legumes and grasses, are excellent cover crops that improve soil fertility and structure. They add organic matter, fix nitrogen, and prevent soil erosion during the off-season.

- **Soil Conditioning**: Annuals can be used to condition the soil before planting perennials. For example, deep-rooted annuals like daikon radishes can break up compacted soil and improve aeration.

4. Pest and Disease Management:

- **Disruption of Pest Cycles**: By changing the types of plants grown in an area each season, annuals can disrupt pest and disease cycles, reducing the likelihood of infestations and outbreaks.

- **Attracting Beneficial Insects**: Many annual flowers and herbs attract beneficial insects that prey on garden pests, enhancing natural pest control.

Planning and Design for Annuals in a Permaculture Garden

1. Garden Layout and Zoning:

- **Zones 1 and 2:** Place annuals in Zones 1 and 2, the areas of the garden closest to the home. These zones are visited frequently, making it convenient to plant, maintain, and harvest annual crops.

- Intercropping and Polycultures: Integrate annuals with perennials in polyculture designs. Intercropping annuals among perennials can optimize space, improve plant health, and increase biodiversity.

2. Succession Planting:

- Continuous Harvest: Plan for succession planting to ensure continuous production throughout the growing season. As one crop is harvested, another can be planted in its place, maintaining a steady supply of fresh produce.

- **Staggered Planting**: Plant annuals in stages to spread out the harvest period. This approach ensures a consistent supply of crops rather than a single, overwhelming harvest.

3. Soil Preparation and Fertility:
- **Compost and Mulch**: Before planting annuals, enrich the soil with compost and mulch to provide essential nutrients and improve soil structure. This preparation creates a healthy growing environment for annual crops.
- **Organic Fertilizers:** Use organic fertilizers, such as compost tea or worm castings, to boost soil fertility and support the fast growth of annuals. Regularly amend the soil to replenish nutrients used by the previous crop.

4. Water Management:
- **Efficient Irrigation**: Implement efficient irrigation techniques, such as drip irrigation or soaker hoses, to provide consistent

moisture to annuals without wasting water. Annuals often have shallow root systems that require regular watering.

- **Mulching:** Apply mulch around annual plants to conserve soil moisture, suppress weeds, and regulate soil temperature. Mulching reduces the need for frequent watering and protects plants from temperature extremes.

5. Companion Planting:

- Mutual Benefits: Practice companion planting by pairing annuals with plants that offer mutual benefits. For example, planting basil near tomatoes can enhance tomato flavor and repel pests.

- **Trap Crops:** Use annuals as trap crops to lure pests away from more valuable plants. For example, planting radishes as a trap crop can protect other brassicas from root maggots.

Best Practices for Integrating Annuals

1. Choosing the Right Annuals:
   - Climate and Season: Select annuals suited to your local climate and growing season. Consider the specific needs of each crop, including temperature, sunlight, and water requirements.
   - **Soil Type:** Choose annuals that thrive in your garden's soil type. Conduct a soil test to determine pH and nutrient levels, and select crops accordingly.

2. Pest and Disease Management:
   - **Diverse Plantings:** Diversify plantings to minimize pest and disease risks. Monocultures are more susceptible to infestations, so mix different types of annuals and perennials.
   - Natural Remedies: Use natural pest control methods, such as neem oil, insecticidal soap, and beneficial insects, to manage pests without harming the garden ecosystem.

3. Seed Saving and Propagation:

- Seed Saving: Save seeds from annual crops to ensure a sustainable supply for future plantings. Select seeds from the healthiest, most productive plants.

- **Transplanting**: Start seeds indoors and transplant seedlings to the garden when conditions are favorable. This method extends the growing season and gives plants a head start.

4. Crop Rotation and Diversity:

- Avoid Replanting Same Family: Avoid planting the same family of annuals in the same location year after year. Crop rotation reduces the risk of soil-borne diseases and nutrient depletion.

- **Diverse Varieties**: Plant a variety of annual crops to promote biodiversity and resilience. Different crops have varying nutrient needs and pest resistance, which benefits the overall garden health.

5. Harvesting and Post-Harvest Management:

- Timely Harvesting: Harvest annual crops at their peak to ensure the best flavor and nutritional value. Regular harvesting also encourages continuous production.

- Storage and Preservation: Preserve excess produce through canning, drying, freezing, or fermenting to extend the availability of home-grown food throughout the year.

Examples of Annuals in Permaculture Gardens

1. Vegetables:

- Tomatoes: Provide abundant yields and can be grown in a variety of climates. Companion plants like basil and marigolds enhance growth and deter pests.

- **Lettuce:** Grows quickly and can be planted multiple times in a season. It thrives in cooler weather and can be interplanted with slower-growing crops.

- **Beans:** Fix nitrogen in the soil, benefiting surrounding plants. Pole beans can be grown on trellises to save space.

2. Herbs:
- Basil: Complements many vegetables and repels pests. It grows well in warm conditions and adds flavor to culinary dishes.
- **Cilantro:** Prefers cooler weather and can be planted in succession for continuous harvest. It attracts beneficial insects.

3. Flowers:
- **Marigolds:** Repel nematodes and attract beneficial insects. They are easy to grow and add color to the garden.
- **Sunflowers:** Provide food for pollinators and birds. Their tall growth habit creates shade for other plants and can serve as a natural trellis.

4. Cover Crops:
   - **Clover:** Fixes nitrogen and improves soil structure. It can be used as a living mulch or green manure.
   - **Buckwheat**: Grows quickly and suppresses weeds. It improves soil fertility and attracts pollinators.

Conclusion

Integrating annuals into a permaculture garden enhances productivity, biodiversity, and resilience. By understanding the benefits of annual plants and implementing strategic planning, gardeners can maximize the advantages of quick-growing crops while maintaining a sustainable and balanced garden system. Annuals provide rapid yields, flexibility, and soil improvement, making them valuable components of a permaculture design. Through thoughtful integration, including companion planting, crop rotation, and efficient water management, annuals can complement perennials to create a dynamic and productive garden that supports both short-term and long-term goals. This holistic approach ensures a steady supply of fresh produce, enhances garden health, and promotes ecological balance.

CHAPTER 5

Garden Layout and Design

Garden Layout and Design in Permaculture Gardening

Designing a permaculture garden involves creating a harmonious, efficient, and sustainable ecosystem that maximizes productivity while minimizing waste and resource consumption. This requires a thoughtful approach that considers natural patterns, site-specific conditions, and permaculture principles. In this comprehensive guide, we'll explore the critical aspects of garden layout and design, providing you with the tools and knowledge to create a thriving permaculture garden.

Understanding Permaculture Design Principles

Permaculture design is rooted in a set of ethical principles and guidelines that aim to create sustainable and self-sufficient ecosystems. Here are some foundational principles to guide your design process:

1. Observe and Interact:
   - Spend time observing your site in all seasons to understand its unique characteristics, including sunlight patterns,

wind direction, water flow, soil types, and existing vegetation.

- Interact with your environment by gardening, walking the site, and engaging with the soil and plants to develop a deep understanding of the land.

2. Catch and Store Energy:

- Design systems that capture and store resources, such as rainwater harvesting systems, solar energy capture, and nutrient cycling through composting.

3. Obtain a Yield:

- Ensure that your garden provides a variety of yields, such as food, fiber, fuel, and medicinal plants. A productive garden contributes to self-sufficiency and sustainability.

4. Apply Self-Regulation and Accept Feedback:

- Monitor your garden regularly and make adjustments based on observations and outcomes. Learn from successes and challenges to improve your design continuously.

5. Use and Value Renewable Resources and Services:
- Prioritize the use of renewable resources, such as solar energy, wind power, and organic matter, to reduce reliance on non-renewable inputs.

6. Design from Patterns to Details:
- Start with broad patterns, such as the overall layout and zoning, and then refine the details, such as plant placement and pathways.

7. Integrate Rather Than Segregate:
- Design for relationships between elements, ensuring that they support and enhance each other. For example, companion planting can improve plant health and yields.

8. Use Small and Slow Solutions:
   - Focus on small, manageable projects that can be expanded over time. This approach allows for gradual learning and adaptation.

9. Use and Value Diversity:
   - Incorporate a variety of plants, animals, and microorganisms to enhance resilience and reduce vulnerability to pests and diseases.

10. Use Edges and Value the Marginal:
   - Utilize the productive potential of edges and transition zones, which often have increased biodiversity and resource availability.

Steps for Designing Your Permaculture Garden Layout

1. Site Assessment:

Before you start designing your garden, conduct a thorough site assessment to gather essential information about your land.

Mapping:
- Create a detailed map of your site, noting key features such as buildings, trees, water sources, slopes, and existing vegetation.

Microclimates:
- Identify microclimates within your garden, such as sunny spots, shady areas, wind-exposed zones, and frost pockets.

Soil Analysis:
- Test your soil to determine its texture, structure, pH, and nutrient levels. This

information will guide your plant selection and soil improvement strategies.

Water Flow:
   - Observe water flow patterns during rain events to identify areas prone to erosion or waterlogging.

2. Zoning:

Permaculture zoning organizes your garden into areas based on the frequency of use and maintenance needs. This approach maximizes efficiency and productivity.

Zone 0:
      - The home or central hub, where activities such as cooking, composting, and tool storage occur.

Zone 1:
   - The area closest to the home, ideal for high-maintenance plants and frequently

harvested crops, such as kitchen gardens and herb beds.

Zone 2:
- A slightly more distant area for lower-maintenance plants, such as larger vegetable plots, berry bushes, and compost bins.

Zone 3:
- Areas for large-scale crops, orchards, and livestock that require less frequent attention.

Zone 4:
- Semi-wild areas for managed forestry, forage, and wildlife habitat.

Zone 5:
- A natural or wild area left undisturbed to encourage biodiversity and provide habitat for wildlife.

3. Sector Analysis:

Sector analysis helps you understand the energy flows and external influences that impact your garden, such as sunlight, wind, water, and wildlife movement.

Energy Flows:
- Analyze how different energies, such as sunlight and wind, interact with your site. Use this information to place elements that either harness or mitigate these energies.

Sector Mapping:
- Overlay sector maps onto your base map to visualize how external factors influence your site.

4. Functional Design Elements:

Incorporate functional design elements to enhance the efficiency and productivity of your garden.

Paths and Access:

- Design paths and access routes to facilitate easy movement and efficient garden management. Ensure paths are wide enough for wheelbarrows and other tools.

Water Management:

- Implement water management features, such as swales, ponds, rain gardens, and greywater systems, to capture, store, and distribute water effectively.

Soil Building:

- Plan for soil-building practices, including composting areas, mulch storage, and green manure crops, to maintain soil fertility and structure.

Structures and Infrastructure:

- Integrate garden structures, such as greenhouses, sheds, trellises, and fences, into the overall design. Ensure they are placed for maximum efficiency and aesthetic appeal.

5. Plant Selection and Placement:

Carefully select and place plants to create a diverse, resilient, and productive garden.

Plant Guilds:
   - Design plant guilds that include complementary species with mutually beneficial relationships. For example, a fruit tree guild might include nitrogen-fixing plants, ground covers, and pollinator-attracting flowers.

Companion Planting:
   - Utilize companion planting strategies to enhance plant health, reduce pests, and optimize resource use.

Layering:
   - Apply the concept of layering, with ground covers, herbaceous plants, shrubs,

understory trees, and canopy trees, to maximize vertical space and increase biodiversity.

6. Succession Planning:

Plan for succession planting to ensure continuous harvests and year-round productivity.

Seasonal Planting:
 - Plan for seasonal succession by rotating crops and using cover crops to maintain soil health.

Long-Term Planning:
 - Consider the long-term growth and development of perennials and trees. Plan for future canopy expansion, root growth, and plant maturity.

7. Maintenance and Monitoring:

Regularly observe and monitor your garden to assess plant health, soil conditions, water management, and pest populations.

Regular Observation:
   - Continuously observe and monitor the garden to make informed adjustments.

Adaptive Management:
   - Be flexible and willing to adapt your design based on feedback from the garden.

Examples of Effective Permaculture Garden Designs

1. Mandala Garden:

Radial Design:
   - A mandala garden features a central focal point with radiating paths and planting beds. This design maximizes space

efficiency and allows for easy access to all areas.

Circular Beds:
- Circular or keyhole-shaped beds facilitate intensive planting and reduce path space, increasing the productive area of the garden.

2. Forest Garden:

Multi-Story Planting:
- A forest garden mimics a natural forest with multiple layers of vegetation, including ground covers, shrubs, understory trees, and canopy trees. This design maximizes vertical space and biodiversity.

Perennial Focus:
- Emphasizes perennial plants that provide long-term yields and require minimal maintenance. The forest garden supports a variety of food, medicinal, and habitat functions.

3. Keyhole Beds:

Efficient Access:
- Keyhole beds have a central path that allows easy access to all parts of the bed without compacting the soil. This design is ideal for intensive planting and small spaces.

Water Conservation:
- The shape of keyhole beds promotes efficient water use by directing water towards plant roots and reducing runoff.

4. Hugelkultur Beds:

Raised Beds:
- Hugelkultur involves creating raised beds with a core of woody debris, which decomposes over time to improve soil fertility and moisture retention.

Water Retention:

- The decomposing wood acts as a sponge, retaining water and reducing the need for irrigation. This design is particularly useful in dry climates.

Best Practices for Garden Layout and Design

1. Start Small and Scale Up:

Begin with a small, manageable area and expand gradually as you gain experience and confidence. This approach allows you to learn from your successes and challenges without becoming overwhelmed.

2. Prioritize Soil Health:

Healthy soil is the foundation of a productive garden. Focus on building and maintaining soil fertility through composting, mulching, and cover cropping.

3. Implement Efficient Water Management:

Water is a critical resource in gardening. Use efficient irrigation techniques, such as drip irrigation and mulching, to conserve water and ensure plants receive adequate moisture.

4. Promote Biodiversity:

Diverse plantings enhance garden resilience and productivity. Incorporate a variety of plants, including perennials, annuals, trees, shrubs, and ground covers, to create a dynamic and balanced ecosystem.

5. Embrace Natural Patterns:

Emulate natural patterns and processes in your garden design. This includes using curved paths, integrating edges and transition zones, and layering plants vertically.

6. Incorporate Multifunctional Elements:

Design garden elements to serve multiple functions, enhancing efficiency and sustainability. For example, a pond can provide water storage, habitat for beneficial insects and amphibians, and a cooling effect on the surrounding microclimate.

7. Foster Beneficial Relationships:

Promote symbiotic relationships between plants, animals, and other elements in your garden. Companion planting, plant guilds, and integrating animals like chickens or bees can enhance productivity and ecosystem health.

8. Utilize Vertical Space:

Maximize the use of vertical space by incorporating trellises, arbors, and vertical gardens. This approach allows for higher yields in smaller areas and can create microclimates beneficial to other plants.

9. Ensure Accessibility:

Design pathways and garden beds to be easily accessible, ensuring that you can reach all areas of the garden for planting, maintenance, and harvesting without compacting the soil.

10. Plan for Aesthetics:

A beautiful garden is not only more enjoyable but can also be more functional. Use color, texture, and plant variety to create visually appealing and inviting spaces that encourage engagement with the garden.

Conclusion

Creating an effective garden layout and design in a permaculture system requires thoughtful planning, observation, and a deep understanding of natural processes. By following permaculture principles, assessing your site, and incorporating multifunctional elements, you can design a garden that is both productive and sustainable. Prioritizing soil health, efficient water management, biodiversity, and the emulation of natural patterns will help create a resilient and thriving garden ecosystem. With careful planning and continuous adaptation, your permaculture garden can provide abundant yields, enhance biodiversity, and contribute to a sustainable and self-sufficient lifestyle.

1. Creating Plant Guilds

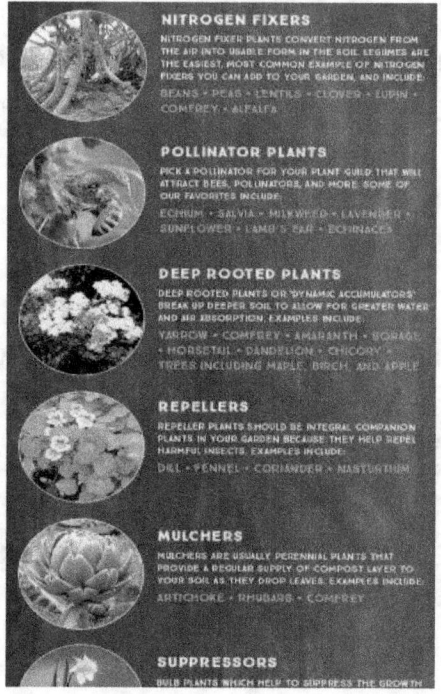

Creating Plant Guilds in Permaculture Gardening

Plant guilds are a cornerstone of permaculture gardening, embodying the principle of beneficial relationships and mimicking natural ecosystems. A plant guild is a group of plants that work together harmoniously, supporting each other's growth, improving soil health, deterring pests, and optimizing resource use. By strategically designing plant guilds, you can create a resilient, productive, and sustainable garden.

Understanding Plant Guilds

1. Definition of Plant Guilds:

A plant guild is a collection of plants that are purposefully grouped to create mutually beneficial relationships. Each plant in the guild serves a specific function that supports the overall health and productivity of the group.

2. Functions of Plant Guilds:

- Nutrient Accumulation: Some plants, known as dynamic accumulators, draw up

nutrients from deep in the soil and make them available to other plants.
- Nitrogen Fixation: Leguminous plants fix atmospheric nitrogen, enriching the soil and providing essential nutrients for companion plants.
- Pest Control: Certain plants repel pests or attract beneficial insects that prey on pests.
- Pollination Support: Flowering plants attract pollinators, enhancing fruit and seed production.
- Ground Cover: Low-growing plants suppress weeds, protect the soil, and retain moisture.
- Mulching and Soil Building: Some plants provide mulch or organic matter that decomposes and enriches the soil.
- Structural Support: Plants like trees and shrubs provide physical support and microclimates for other plants in the guild.

Designing Effective Plant Guilds

1. Core Tree or Central Plant:

Every plant guild typically revolves around a core plant, often a fruit or nut tree, that serves as the primary focus. This core plant provides a canopy and produces yields such as fruits, nuts, or timber.

Example: Apple Tree as a Core Plant

2. Support Species:

Around the core plant, incorporate a variety of support species that fulfill different ecological functions. These may include nitrogen fixers, dynamic accumulators, pest repellents, pollinator attractors, ground covers, and more.

Components of a Typical Plant Guild:

- Nitrogen Fixers: Plants like clover, peas, beans, or alfalfa that enhance soil fertility.

- Dynamic Accumulators: Plants such as comfrey, yarrow, and dandelion that accumulate essential nutrients.
- Pest Repellents: Aromatic herbs like garlic, chives, and marigold that deter pests.
- Pollinator Attractors: Flowers such as echinacea, bee balm, and lavender that attract bees and other pollinators.
- Ground Covers: Plants like creeping thyme, clover, and sweet alyssum that suppress weeds and protect the soil.
- Mulching Plants: Species such as nasturtium and calendula that provide organic matter for mulching.

3. Layering in Guilds:

Guilds often mimic the natural structure of forests, with multiple layers of vegetation. This vertical stratification maximizes space and resource use.

Layers of a Plant Guild:

- Canopy Layer: The tallest plants, usually trees, that form the uppermost layer.
- Understory Layer: Shrubs and smaller trees that grow beneath the canopy.
- Herbaceous Layer: Perennial and annual herbs and vegetables.
- Ground Cover Layer: Low-growing plants that spread across the soil surface.
- Root Layer: Root crops and plants with deep or spreading roots.
- Vine Layer: Climbing plants that utilize vertical space.

4. Creating Symbiotic Relationships:

Ensure that each plant in the guild has a clear role and benefits from the presence of other plants. The relationships should be symbiotic, meaning that the plants support each other's growth and health.

Example of a Symbiotic Plant Guild:

- Core Plant: Apple tree
- Nitrogen Fixers: White clover (ground cover), lupines (flowering plants)
- Dynamic Accumulators: Comfrey (mulch and nutrient accumulator)
- Pest Repellents: Chives (herbaceous layer), marigold (flowering plants)
- Pollinator Attractors: Bee balm (herbaceous layer), yarrow (flowering plants)
- Ground Covers: Creeping thyme (low-growing ground cover)
- Vines: Nasturtium (ground cover and pest repellent)

Implementing Plant Guilds in Your Garden

1. Planning and Mapping:

Start by selecting a core plant suited to your climate and soil conditions. Draw a map of your garden space, marking the location of the core plant and the potential placement of support species.

2. Plant Selection:

Choose support plants that fulfill the necessary functions and are compatible with the core plant. Consider factors such as light, water, and nutrient requirements to ensure harmonious growth.

3. Planting and Establishment:

Plant the core tree first, followed by the support species in their respective layers. Ensure proper spacing to allow for growth and to avoid competition for resources.

4. Maintenance and Observation:

Regularly monitor the guild to assess plant health, growth, and interactions. Adjust the guild composition as needed based on observations and changing conditions.

5. Adaptive Management:

Be flexible and willing to experiment with different plant combinations. Permaculture is an adaptive practice, and learning from successes and challenges will improve your guild design over time.

Examples of Successful Plant Guilds

1. Apple Tree Guild:

- Core Plant: Apple tree
- Nitrogen Fixers: White clover, lupines
- Dynamic Accumulators: Comfrey
- Pest Repellents: Chives, marigold
- Pollinator Attractors: Bee balm, yarrow
- Ground Covers: Creeping thyme
- Vines: Nasturtium

2. Pear Tree Guild:

- Core Plant: Pear tree
- Nitrogen Fixers: Red clover, peas

- Dynamic Accumulators: Borage
- Pest Repellents: Garlic, nasturtium
- Pollinator Attractors: Lavender, coneflower
- Ground Covers: Sweet alyssum
- Vines: Honeysuckle

3. Cherry Tree Guild:

- Core Plant: Cherry tree
- Nitrogen Fixers: Vetch, bush beans
- Dynamic Accumulators: Yarrow
- Pest Repellents: Mint, tansy
- Pollinator Attractors: Phacelia, calendula
- Ground Covers: Creeping Jenny
- Vines: Morning glory

Conclusion

Creating plant guilds is a dynamic and rewarding aspect of permaculture gardening. By understanding the roles and functions of different plants and designing guilds that mimic natural ecosystems, you can enhance the health, productivity, and sustainability of your garden. Plant guilds foster beneficial relationships, improve soil health, support biodiversity, and reduce the need for external inputs. Through thoughtful planning, observation, and adaptation, your plant guilds will thrive, contributing to a resilient and bountiful permaculture garden.

Examples of Plant Guilds

Examples of Plant Guilds

Plant guilds are thoughtfully designed plant communities where each species supports the others, mimicking natural ecosystems.

Apple Tree Guild

The apple tree guild is one of the most popular guilds in temperate climates, designed to maximize the productivity and health of the apple tree while supporting a diverse range of plants.

Core Plant:
- Apple Tree (Malus domestica): The central element providing fruit, shade, and structure.

Support Plants:

Nitrogen Fixers:
- White Clover (Trifolium repens): A low-growing ground cover that fixes nitrogen and suppresses weeds.
- Lupines (Lupinus spp.): Beautiful flowering plants that fix nitrogen and attract pollinators.

Dynamic Accumulators:
- Comfrey (Symphytum officinale): A perennial herb that mines nutrients from deep in the soil and provides mulch material.

Pest Repellents:
- Chives (Allium schoenoprasum): A perennial herb that repels pests like aphids and Japanese beetles.
- Marigold (Tagetes spp.): Annual flowers that repel nematodes and attract beneficial insects.

Pollinator Attractors:
- Bee Balm (Monarda spp.): A perennial herb that attracts bees, butterflies, and hummingbirds.
- Yarrow (Achillea millefolium): A perennial that attracts beneficial insects and accumulates nutrients.

Ground Covers:
- Creeping Thyme (Thymus serpyllum): A low-growing perennial herb that covers the ground, suppresses weeds, and attracts pollinators.

Vines:
- Nasturtium (Tropaeolum majus): An annual vine that repels pests and provides edible flowers and leaves.

Example Layout:
- Center: Apple tree.
- Understory: Comfrey around the base, chives, and marigold interspersed.
- Ground Cover: White clover and creeping thyme.
- Edge: Bee balm, yarrow, and lupines.
- Vine: Nasturtium climbing through the ground cover.

Pear Tree Guild

A pear tree guild follows similar principles to the apple tree guild but uses plants that are specifically beneficial to pear trees.

Core Plant:
- Pear Tree (Pyrus spp.): The central plant providing fruit and structure.

Support Plants:

Nitrogen Fixers:
- Red Clover (Trifolium pratense): A nitrogen-fixing ground cover that also attracts pollinators.
- Peas (Pisum sativum): A nitrogen-fixing annual vine that can climb the tree in early spring.

Dynamic Accumulators:
- Borage (Borago officinalis): An annual herb that accumulates nutrients and attracts pollinators.

Pest Repellents:
- Garlic (Allium sativum): A perennial bulb that repels pests.
- Nasturtium (Tropaeolum majus): Repels pests and attracts beneficial insects.

Pollinator Attractors:
- Lavender (Lavandula spp.): A perennial herb that attracts pollinators and repels pests.
- Coneflower (Echinacea spp.): A perennial that attracts pollinators and beneficial insects.

Ground Covers:
- Sweet Alyssum (Lobularia maritima): A low-growing annual that attracts beneficial insects and covers the ground.

Vines:
- Honeysuckle (Lonicera spp.): A perennial vine that attracts pollinators and provides edible flowers.

Example Layout:
- Center: Pear tree.
- Understory: Borage around the base, garlic, and nasturtium interspersed.
- Ground Cover: Red clover and sweet alyssum.

- Edge: Lavender, coneflower, and peas.
- Vine: Honeysuckle climbing through the lower branches.

Cherry Tree Guild

Cherry trees can benefit from a well-designed guild that supports their specific needs, particularly in terms of pest control and pollination.

Core Plant:
- Cherry Tree (Prunus spp.): The central plant providing fruit and structure.

Support Plants:

Nitrogen Fixers:
- Vetch (Vicia spp.): A nitrogen-fixing ground cover.
- Bush Beans (Phaseolus vulgaris): Nitrogen-fixing annual plants that provide a food yield.

Dynamic Accumulators:
- Yarrow (Achillea millefolium): Accumulates nutrients and attracts beneficial insects.

Pest Repellents:
- Mint (Mentha spp.): A perennial herb that repels pests and spreads quickly.
- Tansy (Tanacetum vulgare): A perennial that repels a variety of pests.

Pollinator Attractors:
- Phacelia (Phacelia spp.): A flowering annual that attracts pollinators and beneficial insects.
- Calendula (Calendula officinalis): An annual flower that attracts pollinators and deters pests.

Ground Covers:
- Creeping Jenny (Lysimachia nummularia): A low-growing perennial that covers the ground and retains moisture.

Vines:
- Morning Glory (Ipomoea spp.): An annual vine that attracts pollinators with its flowers.

Example Layout:
- Center: Cherry tree.
- Understory: Yarrow around the base, mint, and tansy interspersed.
- Ground Cover: Vetch and creeping Jenny.
- Edge: Phacelia, calendula, and bush beans.
- Vine: Morning glory climbing through the lower branches.

Walnut Tree Guild

Walnut trees produce juglone, a substance toxic to many plants. Designing a guild around a walnut tree requires selecting juglone-tolerant species.

Core Plant:
- Walnut Tree (Juglans spp.): The central plant providing nuts and structure.

Support Plants:

Nitrogen Fixers:
- Black Locust (Robinia pseudoacacia): A nitrogen-fixing tree that tolerates juglone and can provide additional shade and structure.
- Clover (Trifolium spp.): A ground cover that fixes nitrogen and tolerates juglone.

Dynamic Accumulators:
- Comfrey (Symphytum officinale): Tolerates juglone and provides mulch material.

Pest Repellents:
- Garlic (Allium sativum): Repels pests and tolerates juglone.
- Chives (Allium schoenoprasum): Repels pests and tolerates juglone.

Pollinator Attractors:
- Bee Balm (Monarda spp.): Attracts pollinators and tolerates juglone.

- Black-eyed Susan (Rudbeckia hirta): Attracts pollinators and tolerates juglone.

Ground Covers:
- Sweet Woodruff (Galium odoratum): A low-growing plant that tolerates juglone and covers the ground.

Example Layout:
- Center: Walnut tree.
- Understory: Black locust trees and comfrey around the base, garlic, and chives interspersed.
- Ground Cover: Clover and sweet woodruff.
- Edge: Bee balm and black-eyed Susan.

Conclusion

Plant guilds are integral to permaculture gardening, promoting mutual benefits among plants and creating resilient ecosystems. Each guild is designed to support the central plant, typically a tree, through a variety of supporting species that enhance soil health, attract pollinators, repel pests, and improve overall productivity. By understanding the specific needs and compatibility of different plants, you can design effective and sustainable plant guilds tailored to your garden's unique conditions. These examples provide a foundation for creating your own plant guilds, encouraging biodiversity and fostering a healthy, thriving garden.

Benefits of Polycultures

Benefits of Polycultures in Permaculture Gardening

Polycultures, the practice of growing multiple crop species together in a shared space, are a key element of permaculture gardening. This method stands in contrast to monocultures, where a single crop is cultivated extensively. Polycultures offer numerous benefits, including enhanced biodiversity, pest control, soil health, resilience, and productivity.

Enhanced Biodiversity

1. Ecological Balance:

Polycultures mimic natural ecosystems by incorporating a variety of plant species, creating a balanced and diverse environment. This diversity supports a range of organisms, including beneficial insects, birds, and soil microorganisms, contributing to a healthier ecosystem.

2. Genetic Diversity:

Cultivating different plant species enhances genetic diversity within the garden. This diversity reduces the risk of widespread disease and pest outbreaks, as not all species are equally susceptible to the same threats.

3. Habitat Creation:

A diverse planting strategy creates habitats for various wildlife, including pollinators and predatory insects. These organisms help control pest populations and promote pollination, leading to better yields and a healthier garden.

Pest Control

1. Natural Pest Management:

Polycultures reduce the prevalence of pests by breaking up the uniformity of plantings that pests often exploit in monocultures. Pests are less likely to spread rapidly in a diverse planting system because their preferred host plants are dispersed.

2. Attraction of Beneficial Insects:

Certain plants in a polyculture can attract beneficial insects that prey on pests. For example, flowering plants such as marigolds and calendula attract ladybugs and lacewings, which feed on aphids and other harmful insects.

3. Repellent Plants:

Incorporating pest-repellent plants within a polyculture can help protect other crops. Aromatic herbs like basil, mint, and rosemary can deter pests through their strong scents, reducing the need for chemical interventions.

Soil Health

1. Nutrient Cycling:

Different plants have varying nutrient requirements and root structures, which helps in efficient nutrient cycling. Deep-rooted plants can access nutrients from deeper soil layers and bring them to the surface, making them available to shallow-rooted plants.

2. Soil Structure Improvement:

A variety of root systems within a polyculture improves soil structure by reducing compaction and enhancing aeration. This results in better water infiltration and retention, supporting healthier plant growth.

3. Erosion Control:

The varied canopy cover and root structures in polycultures help protect the soil from erosion by wind and water. Ground-cover plants, in particular, play a significant role in preventing soil erosion by shielding the soil surface.

Resilience and Stability

1. Disease Resistance:

Polycultures reduce the spread of diseases by creating a more complex environment where pathogens struggle to locate and infect their preferred host plants. This reduces the overall disease pressure within the garden.

2. Climate Resilience:

A diverse planting system can better withstand extreme weather conditions. Different species respond differently to temperature fluctuations, drought, and heavy rainfall, ensuring that some plants will thrive even under adverse conditions.

3. Risk Management:

By growing multiple crops, gardeners spread the risk of crop failure. If one crop fails due to pests, disease, or weather, other crops may still succeed, ensuring some level of harvest.

Productivity and Resource Use Efficiency

1. Maximized Yield:

Polycultures often lead to higher overall yields compared to monocultures. Different plants utilize resources (light, water, nutrients) at varying times and depths, reducing competition and making better use of available resources.

2. Intercropping Benefits:

Intercropping, a form of polyculture, allows for the strategic pairing of plants to enhance growth. For example, the "Three Sisters" method used by Native Americans involves growing corn, beans, and squash together. Corn provides a structure for beans to climb, beans fix nitrogen in the soil, and squash spreads along the ground, suppressing weeds and retaining moisture.

3. Continuous Harvest:

Planting a variety of species with different maturation times ensures a continuous harvest throughout the growing season. This provides a steady supply of fresh produce and spreads labor needs over time.

Environmental and Economic Benefits

1. Reduced Chemical Inputs:

The natural pest and disease control benefits of polycultures reduce the need for chemical pesticides and fertilizers. This lowers the environmental impact of gardening and promotes a healthier ecosystem.

2. Economic Savings:

Reduced reliance on chemical inputs and improved soil health lead to long-term economic savings for gardeners. Additionally, diverse crops can provide a variety of products for sale or personal use, enhancing economic stability.

3. Enhanced Ecosystem Services:

Polycultures contribute to various ecosystem services, including pollination, soil formation, nutrient cycling, and water regulation. These services are crucial for the sustainability of agricultural systems and the broader environment.

Examples of Successful Polycultures

1. Three Sisters (Corn, Beans, and Squash):

- **Corn**: Provides a natural trellis for beans.

- Beans: Fix nitrogen in the soil, benefiting all plants.
- **Squash:** Covers the ground, suppressing weeds and retaining soil moisture.

2. Agroforestry Systems:

Combining trees, shrubs, and crops in the same area enhances biodiversity, improves soil health, and increases resilience. Examples include alley cropping and silvopasture.

3. Mixed Vegetable Gardens:

Planting a variety of vegetables, herbs, and flowers together supports pest control, pollination, and soil health. For instance, planting tomatoes with basil and marigolds can deter pests and improve tomato growth.

Conclusion

Polycultures are a fundamental practice in permaculture gardening, offering a multitude of benefits that contribute to a sustainable, resilient, and productive garden. By enhancing biodiversity, naturally managing pests, improving soil health, increasing resilience to climate extremes, and maximizing productivity, polycultures create a balanced and thriving ecosystem. Embracing polycultures in gardening not only supports the health of the garden but also promotes environmental sustainability and economic stability. Through careful planning and strategic planting, gardeners can harness the power of polycultures to achieve a more sustainable and fruitful gardening experience.

2. Edible Landscapes

Edible Landscapes

Edible landscapes represent the harmonious integration of ornamental and edible plants within a garden or landscape design. This concept transforms traditional ornamental gardens into spaces that are both aesthetically pleasing and productive. Edible landscapes offer numerous benefits, including sustainable food production, biodiversity enhancement, cost savings, and environmental sustainability. This guide explores the principles, benefits, design strategies, and plant choices for creating a thriving edible landscape.

Principles of Edible Landscaping

1. Integration of Aesthetics and Functionality:

The core principle of edible landscaping is to combine beauty with utility. This involves designing spaces that are visually appealing while also providing edible produce such as fruits, vegetables, herbs, and nuts.

2. Diverse Plant Selection:

Diversity is key to a successful edible landscape. Incorporating a wide range of plants ensures a continuous harvest, supports biodiversity, and reduces the risk of pest and disease outbreaks.

3. Sustainable Practices:

Edible landscapes prioritize sustainable gardening practices, including organic soil amendments, composting, water conservation, and integrated pest management (IPM). These practices enhance soil health, reduce environmental impact, and promote ecological balance.

4. Maximizing Space and Resources:

Efficient use of space and resources is essential. Vertical gardening, companion planting, and succession planting are techniques used to maximize the productivity of the landscape.

5. Perennial Emphasis:

Focusing on perennial plants reduces maintenance efforts and ensures long-term productivity. Perennials such as fruit trees, berry bushes, and perennial herbs provide reliable harvests year after year.

Benefits of Edible Landscaping

1. Sustainable Food Production:

Edible landscapes contribute to food security by providing fresh, organic produce. This reduces reliance on commercial agriculture and decreases the carbon footprint associated with food transportation.

2. Enhanced Biodiversity:

A diverse mix of edible and ornamental plants supports a wide range of pollinators, beneficial insects, birds, and other wildlife. This biodiversity creates a balanced ecosystem and enhances the health of the garden.

3. Environmental Sustainability:

Edible landscapes promote sustainable gardening practices, including organic soil management, water conservation, and reduced use of synthetic chemicals. These practices improve soil health, conserve water, and protect local ecosystems.

4. Economic Savings:

Growing your own food reduces grocery bills and provides access to high-quality, organic produce. Additionally, edible landscapes can increase property values by enhancing the beauty and functionality of the garden.

5. Health and Well-being:

Gardening provides physical exercise, reduces stress, and encourages outdoor activity. Consuming homegrown produce also improves nutrition and promotes a healthier lifestyle.

Design Strategies for Edible Landscapes

1. Site Assessment:

Conduct a thorough assessment of the site, including soil quality, sunlight exposure, water availability, and existing vegetation. This information will guide plant selection and placement.

2. Zoning and Layout:

Divide the garden into zones based on plant needs and garden functions. For example, place high-maintenance plants near the home for easy access and locate larger, low-maintenance perennials farther away.

3. Layering and Stacking:

Use vertical space by incorporating different plant layers, such as ground covers, herbs, shrubs, and trees. This mimics natural ecosystems and maximizes space utilization.

4. Companion Planting:

Select plant combinations that benefit each other. Companion planting improves pest control, enhances soil health, and increases yields. Examples include planting tomatoes with basil or carrots with onions.

5. Seasonal Planning:

Plan for year-round production by including a mix of cool-season and warm-season crops. Succession planting ensures continuous harvests throughout the growing season.

6. Aesthetic Integration:

Incorporate ornamental elements to enhance the visual appeal of the garden. Use edible flowers, attractive vegetable varieties, and decorative structures such as trellises and arbors.

Plant Choices for Edible Landscapes

1. Fruit Trees:

- Apple (Malus domestica): Offers a variety of cultivars with different flavors and ripening times.
- Pear (Pyrus spp.): Hardy and productive, with both Asian and European varieties.
- Cherry (Prunus spp.): Provides sweet or tart fruit and attractive spring blossoms.
- Peach (Prunus persica): Produces juicy fruit and adds ornamental value with its flowers.

2. Berry Bushes:

- Blueberry (Vaccinium spp.): Requires acidic soil and produces antioxidant-rich berries.

- Raspberry (Rubus idaeus): Thrives in well-drained soil and provides a summer harvest.
- Blackberry (Rubus fruticosus): Grows vigorously and produces abundant fruit.
- Currant (Ribes spp.): Offers tart berries for jams and jellies.

3. Vegetables:

- Tomatoes (Solanum lycopersicum): Versatile and available in numerous varieties.
- Lettuce (Lactuca sativa): Fast-growing and ideal for continuous harvests.
- Peppers (Capsicum spp.): Adds color and flavor, with sweet and hot varieties.
- Carrots (Daucus carota): Grows well in loose, well-drained soil.

4. Herbs:

- Basil (Ocimum basilicum): Aromatic and essential for culinary use.

- Rosemary (Salvia rosmarinus): Woody perennial with fragrant leaves.
- Thyme (Thymus vulgaris): Low-growing and drought-tolerant.
- Mint (Mentha spp.): Spreads easily and adds flavor to beverages and dishes.

5. Edible Flowers:

- Nasturtium (Tropaeolum majus): Adds color and peppery flavor to salads.
- Calendula (Calendula officinalis): Bright flowers and edible petals.
- Borage (Borago officinalis): Star-shaped flowers with a cucumber flavor.
- Pansy (Viola tricolor): Colorful blooms that are mild in flavor.

6. Perennial Vegetables:

- Asparagus (Asparagus officinalis): Long-lived and productive after establishment.

- Rhubarb (Rheum rhabarbarum): Grows well in cool climates and provides tart stalks.
- Jerusalem Artichoke (Helianthus tuberosus): Produces edible tubers and attractive flowers.

Water Management in Edible Landscapes

1. Rainwater Harvesting:

Collect rainwater using barrels or cisterns to irrigate the garden. This reduces reliance on municipal water and ensures a sustainable water supply.

2. Efficient Irrigation:

Implement drip irrigation or soaker hoses to deliver water directly to plant roots. This minimizes water loss through evaporation and runoff.

3. Mulching:

Use organic mulches such as straw, wood chips, or leaves to retain soil moisture, suppress weeds, and improve soil structure.

4. Water-wise Planting:

Select drought-tolerant plants for areas with limited water availability. Group plants with similar water needs together to optimize irrigation.

Maintenance and Care

1. Soil Health:

Regularly amend soil with organic matter, such as compost or well-rotted manure, to maintain fertility and structure. Conduct soil tests to monitor nutrient levels and pH.

2. Pruning and Training:

Prune fruit trees and berry bushes to maintain shape, remove dead or diseased wood, and encourage productive growth. Train vines and climbing plants on trellises to optimize space.

3. Pest and Disease Management:

Practice integrated pest management (IPM) by encouraging beneficial insects, using physical barriers, and applying organic treatments when necessary. Monitor plants regularly for signs of pests and diseases.

4. Crop Rotation:

Rotate annual crops to different areas of the garden each year to prevent soil depletion and reduce the buildup of pests and diseases.

5. Harvesting:

Harvest produce regularly to encourage continued production and prevent over-ripening. Use appropriate harvesting techniques to avoid damaging plants.

Conclusion

Edible landscapes offer a sustainable and productive approach to gardening that combines beauty with functionality. By integrating a diverse array of edible and ornamental plants, gardeners can create spaces that provide fresh produce, enhance biodiversity, and promote environmental sustainability. With careful planning, thoughtful design, and sustainable practices, an edible landscape can become a thriving, multifunctional garden that nourishes both the body and the soul.

Integrating Food Forests

Introduction to Food Forests

A food forest is a sustainable, productive, and low-maintenance agricultural system modeled after natural forests and woodlands. It incorporates a diverse range of edible plants organized into different layers, mimicking the natural ecosystem's structure and processes. This approach maximizes space, enhances biodiversity, improves soil health, and creates a resilient and self-sustaining environment.

Food forests can be integrated into various scales, from small backyard gardens to larger community or farm settings. They offer numerous benefits, including food security, environmental sustainability, and aesthetic appeal.

Principles of Food Forest Design

1. Mimicking Natural Ecosystems:
   - Food forests replicate the structure and function of natural forests, incorporating diverse plant species arranged in multiple layers.

2. Diversity and Resilience:
   - A wide variety of plants enhances biodiversity, reduces pest and disease pressure, and creates a more resilient ecosystem.

3. Perennial Emphasis:

- Focusing on perennial plants minimizes maintenance and ensures long-term productivity. Perennials require less replanting and can offer harvests year after year.

4. Sustainable Practices:

- Organic soil management, water conservation, and integrated pest management (IPM) are key to maintaining a healthy and productive food forest.

5. Closed-Loop Systems:

- Food forests aim to create closed-loop systems where waste products (e.g., fallen leaves, plant residues) are recycled back into the ecosystem, enhancing soil fertility and structure.

Layers of a Food Forest

1. Canopy Layer:
   - Tall fruit and nut trees form the uppermost layer, providing shade, wind protection, and a habitat for wildlife. Examples include apple, pear, chestnut, and walnut trees.

2. Low-Tree Layer (Sub-Canopy):
   - Smaller fruit trees and shrubs occupy this layer. Examples include dwarf apple trees, hazelnuts, and persimmons.

3. Shrub Layer:
   - Berry bushes and other edible shrubs form this layer. Examples include blueberries, raspberries, currants, and gooseberries.

4. Herbaceous Layer:
   - This layer includes herbs, vegetables, and perennial flowers. Examples are mint, oregano, rhubarb, and asparagus.

5. Ground Cover Layer:
- Low-growing plants that protect the soil, suppress weeds, and retain moisture. Examples include strawberries, creeping thyme, and clover.

6. Root Layer:
- Root vegetables and tubers grow in this layer, contributing to soil health and providing edible produce. Examples are carrots, potatoes, garlic, and Jerusalem artichokes.

7. Vertical Layer:
- Vining plants that climb up trees, trellises, or other supports. Examples include grapes, kiwis, and climbing beans.

Steps to Integrate a Food Forest

1. Site Assessment:
- Conduct a thorough site analysis, considering factors such as soil quality,

sunlight exposure, water availability, and existing vegetation.

- Identify microclimates within the site to determine the best locations for different plant species.

2. Design and Planning:

- Create a detailed design plan that outlines the placement of each plant layer. Consider plant spacing, companion planting, and the overall layout.

- Include pathways and access points to facilitate maintenance and harvesting.

3. Soil Preparation:

- Improve soil quality by adding organic matter such as compost, manure, and mulch. This enhances soil structure, fertility, and moisture retention.

- Conduct soil tests to determine pH and nutrient levels, and amend as necessary.

4. Plant Selection:
- Choose a diverse range of plants that are well-suited to the local climate and soil conditions. Prioritize native and adaptable species to ensure resilience and minimal maintenance.
- Select plants for each layer, considering their growth habits, light requirements, and companion planting benefits.

5. Planting and Establishment:
- Begin by planting the canopy and sub-canopy layers, followed by the shrub, herbaceous, ground cover, root, and vertical layers.
- Mulch around plants to retain moisture, suppress weeds, and improve soil health.

6. Water Management:
- Implement efficient irrigation methods, such as drip systems or soaker hoses, to provide adequate water to plants, especially during establishment.

- Incorporate water conservation techniques like rainwater harvesting and swales to capture and retain water.

7. Ongoing Maintenance:
- Regularly monitor plant health, soil moisture, and nutrient levels. Adjust irrigation and fertilization practices as needed.
- Prune trees and shrubs to maintain their shape, improve air circulation, and enhance productivity.
- Use organic pest and disease management techniques to protect plants and maintain ecological balance.

Benefits of Food Forests

1. Food Security:
- Food forests provide a reliable source of fresh, organic produce, contributing to food security and self-sufficiency.

2. Environmental Sustainability:

- By mimicking natural ecosystems, food forests enhance biodiversity, improve soil health, and reduce the need for synthetic inputs.

3. Resilience:

- The diverse plant species in food forests create a resilient system that can better withstand pests, diseases, and environmental stresses.

4. Carbon Sequestration:

- Trees and perennial plants in food forests sequester carbon, helping to mitigate climate change.

5. Habitat Creation:

- Food forests provide habitat for a wide range of wildlife, including pollinators, birds, and beneficial insects.

6. Low Maintenance:

- Once established, food forests require less maintenance compared to traditional annual crop systems, reducing labor and resource inputs.

7. Educational and Community Value:

- Food forests serve as educational tools, demonstrating sustainable agriculture practices and fostering community engagement and collaboration.

Examples of Successful Food Forests

1. Beacon Food Forest (Seattle, USA):

- A community-driven project that transformed a public park into a thriving food forest, providing fresh produce to the local community and serving as an educational resource.

2. **Martin Crawford's Forest Garden (Devon, UK):**
 - A pioneering example of temperate food forestry, showcasing a diverse range of edible plants and sustainable gardening practices.

3. **The Permaculture Institute's Food Forest (Santa Fe, USA):**
 - A demonstration site that integrates permaculture principles with food forestry, promoting sustainable land management and ecological design.

Conclusion

Integrating food forests into landscapes offers a sustainable, productive, and aesthetically pleasing approach to agriculture and gardening. By mimicking natural ecosystems, food forests enhance biodiversity, improve soil health, and create resilient, self-sustaining environments. With thoughtful design, careful plant selection, and sustainable practices, food forests can provide abundant harvests, environmental benefits, and a beautiful, functional landscape. Whether on a small scale in a backyard or a larger community setting, food forests represent a promising solution for sustainable food production and ecological harmony.

CHAPTER 6

Sustainable Practices

Introduction to Food Forests

A food forest is a sustainable, productive, and low-maintenance agricultural system modeled after natural forests and woodlands. It incorporates a diverse range of edible plants organized into different layers, mimicking the natural ecosystem's structure and processes. This approach maximizes space, enhances biodiversity, improves soil health, and creates a resilient and self-sustaining environment.

Food forests can be integrated into various scales, from small backyard gardens to larger community or farm settings. They offer numerous benefits, including food security, environmental sustainability, and aesthetic appeal.

Principles of Food Forest Design

1. Mimicking Natural Ecosystems:
- Food forests replicate the structure and function of natural forests, incorporating diverse plant species arranged in multiple layers.

2. Diversity and Resilience:
- A wide variety of plants enhances biodiversity, reduces pest and disease pressure, and creates a more resilient ecosystem.

3. Perennial Emphasis:
- Focusing on perennial plants minimizes maintenance and ensures long-term productivity. Perennials require less replanting and can offer harvests year after year.

4. Sustainable Practices:
- Organic soil management, water conservation, and integrated pest

management (IPM) are key to maintaining a healthy and productive food forest.

5. Closed-Loop Systems:
- Food forests aim to create closed-loop systems where waste products (e.g., fallen leaves, plant residues) are recycled back into the ecosystem, enhancing soil fertility and structure.

Layers of a Food Forest

1. Canopy Layer:
- Tall fruit and nut trees form the uppermost layer, providing shade, wind protection, and a habitat for wildlife. Examples include apple, pear, chestnut, and walnut trees.

2. Low-Tree Layer (Sub-Canopy):
- Smaller fruit trees and shrubs occupy this layer. Examples include dwarf apple trees, hazelnuts, and persimmons.

3. Shrub Layer:

- Berry bushes and other edible shrubs form this layer. Examples include blueberries, raspberries, currants, and gooseberries.

4. Herbaceous Layer:

- This layer includes herbs, vegetables, and perennial flowers. Examples are mint, oregano, rhubarb, and asparagus.

5. Ground Cover Layer:

- Low-growing plants that protect the soil, suppress weeds, and retain moisture. Examples include strawberries, creeping thyme, and clover.

6. Root Layer:

- Root vegetables and tubers grow in this layer, contributing to soil health and providing edible produce. Examples are carrots, potatoes, garlic, and Jerusalem artichokes.

7. Vertical Layer:
   - Vining plants that climb up trees, trellises, or other supports. Examples include grapes, kiwis, and climbing beans.

Steps to Integrate a Food Forest

1. Site Assessment:
   - Conduct a thorough site analysis, considering factors such as soil quality, sunlight exposure, water availability, and existing vegetation.
   - Identify microclimates within the site to determine the best locations for different plant species.

2. Design and Planning:
   - Create a detailed design plan that outlines the placement of each plant layer. Consider plant spacing, companion planting, and the overall layout.
   - Include pathways and access points to facilitate maintenance and harvesting.

3. Soil Preparation:

- Improve soil quality by adding organic matter such as compost, manure, and mulch. This enhances soil structure, fertility, and moisture retention.
- Conduct soil tests to determine pH and nutrient levels, and amend as necessary.

4. Plant Selection:

- Choose a diverse range of plants that are well-suited to the local climate and soil conditions. Prioritize native and adaptable species to ensure resilience and minimal maintenance.
- Select plants for each layer, considering their growth habits, light requirements, and companion planting benefits.

5. Planting and Establishment:

- Begin by planting the canopy and sub-canopy layers, followed by the shrub, herbaceous, ground cover, root, and vertical layers.

- Mulch around plants to retain moisture, suppress weeds, and improve soil health.

6. Water Management:
- Implement efficient irrigation methods, such as drip systems or soaker hoses, to provide adequate water to plants, especially during establishment.
- Incorporate water conservation techniques like rainwater harvesting and swales to capture and retain water.

7. Ongoing Maintenance:
- Regularly monitor plant health, soil moisture, and nutrient levels. Adjust irrigation and fertilization practices as needed.
- Prune trees and shrubs to maintain their shape, improve air circulation, and enhance productivity.
- Use organic pest and disease management techniques to protect plants and maintain ecological balance.

Benefits of Food Forests

1. Food Security:
- Food forests provide a reliable source of fresh, organic produce, contributing to food security and self-sufficiency.

2. Environmental Sustainability:
- By mimicking natural ecosystems, food forests enhance biodiversity, improve soil health, and reduce the need for synthetic inputs.

3. Resilience:
- The diverse plant species in food forests create a resilient system that can better withstand pests, diseases, and environmental stresses.

4. Carbon Sequestration:
- Trees and perennial plants in food forests sequester carbon, helping to mitigate climate change.

5. Habitat Creation:
- Food forests provide habitat for a wide range of wildlife, including pollinators, birds, and beneficial insects.

6. Low Maintenance:
- Once established, food forests require less maintenance compared to traditional annual crop systems, reducing labor and resource inputs.

7. Educational and Community Value:
- Food forests serve as educational tools, demonstrating sustainable agriculture practices and fostering community engagement and collaboration.

Examples of Successful Food Forests

1. Beacon Food Forest (Seattle, USA):
- A community-driven project that transformed a public park into a thriving food forest, providing fresh produce to the

local community and serving as an educational resource.

2. Martin Crawford's Forest Garden (Devon, UK):

- A pioneering example of temperate food forestry, showcasing a diverse range of edible plants and sustainable gardening practices.

3. The Permaculture Institute's Food Forest (Santa Fe, USA):

- A demonstration site that integrates permaculture principles with food forestry, promoting sustainable land management and ecological design.

Conclusion

Integrating food forests into landscapes offers a sustainable, productive, and aesthetically pleasing approach to agriculture and gardening. By mimicking natural ecosystems, food forests enhance biodiversity, improve soil health, and create resilient, self-sustaining environments. With thoughtful design, careful plant selection, and sustainable practices, food forests can provide abundant harvests, environmental benefits, and a beautiful, functional landscape. Whether on a small scale in a backyard or a larger community setting, food forests represent a promising solution for sustainable food production and ecological harmony.

1. Natural Pest Control

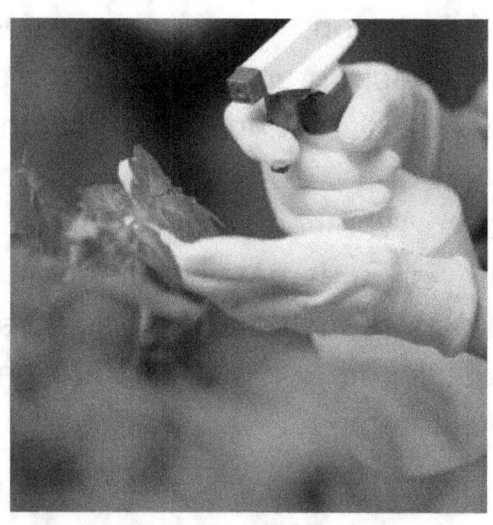

Introduction to Food Forests

A food forest is a sustainable, productive, and low-maintenance agricultural system modeled after natural forests and woodlands. It incorporates a diverse range of edible plants organized into different layers, mimicking the natural ecosystem's structure and processes. This approach maximizes space, enhances biodiversity, improves soil health, and creates a resilient and self-sustaining environment.

Food forests can be integrated into various scales, from small backyard gardens to larger community or farm settings. They offer numerous benefits, including food security, environmental sustainability, and aesthetic appeal.

Principles of Food Forest Design

1. Mimicking Natural Ecosystems:
   - Food forests replicate the structure and function of natural forests, incorporating diverse plant species arranged in multiple layers.

2. Diversity and Resilience:
   - A wide variety of plants enhances biodiversity, reduces pest and disease pressure, and creates a more resilient ecosystem.

3. Perennial Emphasis:
- Focusing on perennial plants minimizes maintenance and ensures long-term productivity. Perennials require less replanting and can offer harvests year after year.

4. Sustainable Practices:
- Organic soil management, water conservation, and integrated pest management (IPM) are key to maintaining a healthy and productive food forest.

5. Closed-Loop Systems:
- Food forests aim to create closed-loop systems where waste products (e.g., fallen leaves, plant residues) are recycled back into the ecosystem, enhancing soil fertility and structure.

Layers of a Food Forest

1. Canopy Layer:
- Tall fruit and nut trees form the uppermost layer, providing shade, wind protection, and a habitat for wildlife. Examples include apple, pear, chestnut, and walnut trees.

2. Low-Tree Layer (Sub-Canopy):
- Smaller fruit trees and shrubs occupy this layer. Examples include dwarf apple trees, hazelnuts, and persimmons.

3. Shrub Layer:
- Berry bushes and other edible shrubs form this layer. Examples include blueberries, raspberries, currants, and gooseberries.

4. Herbaceous Layer:
- This layer includes herbs, vegetables, and perennial flowers. Examples are mint, oregano, rhubarb, and asparagus.

5. Ground Cover Layer:
- Low-growing plants that protect the soil, suppress weeds, and retain moisture. Examples include strawberries, creeping thyme, and clover.

6. Root Layer:
- Root vegetables and tubers grow in this layer, contributing to soil health and providing edible produce. Examples are carrots, potatoes, garlic, and Jerusalem artichokes.

7. Vertical Layer:
- Vining plants that climb up trees, trellises, or other supports. Examples include grapes, kiwis, and climbing beans.

Steps to Integrate a Food Forest

1. Site Assessment:
- Conduct a thorough site analysis, considering factors such as soil quality,

sunlight exposure, water availability, and existing vegetation.
- Identify microclimates within the site to determine the best locations for different plant species.

2. Design and Planning:
- Create a detailed design plan that outlines the placement of each plant layer. Consider plant spacing, companion planting, and the overall layout.
- Include pathways and access points to facilitate maintenance and harvesting.

3. Soil Preparation:
- Improve soil quality by adding organic matter such as compost, manure, and mulch. This enhances soil structure, fertility, and moisture retention.
- Conduct soil tests to determine pH and nutrient levels, and amend as necessary.

4. Plant Selection:
- Choose a diverse range of plants that are well-suited to the local climate and soil conditions. Prioritize native and adaptable species to ensure resilience and minimal maintenance.
- Select plants for each layer, considering their growth habits, light requirements, and companion planting benefits.

5. Planting and Establishment:
- Begin by planting the canopy and sub-canopy layers, followed by the shrub, herbaceous, ground cover, root, and vertical layers.
- Mulch around plants to retain moisture, suppress weeds, and improve soil health.

6. Water Management:
- Implement efficient irrigation methods, such as drip systems or soaker hoses, to provide adequate water to plants, especially during establishment.

- Incorporate water conservation techniques like rainwater harvesting and swales to capture and retain water.

7. Ongoing Maintenance:
- Regularly monitor plant health, soil moisture, and nutrient levels. Adjust irrigation and fertilization practices as needed.
- Prune trees and shrubs to maintain their shape, improve air circulation, and enhance productivity.
- Use organic pest and disease management techniques to protect plants and maintain ecological balance.

Benefits of Food Forests

1. Food Security:
- Food forests provide a reliable source of fresh, organic produce, contributing to food security and self-sufficiency.

2. Environmental Sustainability:
- By mimicking natural ecosystems, food forests enhance biodiversity, improve soil health, and reduce the need for synthetic inputs.

3. Resilience:
- The diverse plant species in food forests create a resilient system that can better withstand pests, diseases, and environmental stresses.

4. Carbon Sequestration:
- Trees and perennial plants in food forests sequester carbon, helping to mitigate climate change.

5. Habitat Creation:
- Food forests provide habitat for a wide range of wildlife, including pollinators, birds, and beneficial insects.

6. Low Maintenance:

- Once established, food forests require less maintenance compared to traditional annual crop systems, reducing labor and resource inputs.

7. Educational and Community Value:

- Food forests serve as educational tools, demonstrating sustainable agriculture practices and fostering community engagement and collaboration.

Examples of Successful Food Forests

1. Beacon Food Forest (Seattle, USA):
    - A community-driven project that transformed a public park into a thriving food forest, providing fresh produce to the local community and serving as an educational resource.

2. Martin Crawford's Forest Garden (Devon, UK):

- A pioneering example of temperate food forestry, showcasing a diverse range of edible plants and sustainable gardening practices.

3. The Permaculture Institute's Food Forest (Santa Fe, USA):

- A demonstration site that integrates permaculture principles with food forestry, promoting sustainable land management and ecological design.

Conclusion

Integrating food forests into landscapes offers a sustainable, productive, and aesthetically pleasing approach to agriculture and gardening. By mimicking natural ecosystems, food forests enhance biodiversity, improve soil health, and create resilient, self-sustaining environments. With thoughtful design, careful plant selection, and sustainable practices, food forests can provide abundant harvests, environmental benefits, and a beautiful, functional landscape. Whether on a small scale in a backyard or a larger community setting, food forests represent a promising solution for sustainable food production and ecological harmony.

Encouraging Beneficial Insects

Introduction to Beneficial Insects

Beneficial insects play a crucial role in maintaining the health and balance of your garden ecosystem. These insects contribute to pest control, pollination, soil health, and overall biodiversity. Encouraging beneficial insects is a key aspect of sustainable gardening and permaculture, as it reduces the need for chemical interventions and fosters a resilient environment.

Types of Beneficial Insects

1. Pollinators:
- **Bees (Apis spp**.): Essential for pollinating a wide range of plants, ensuring fruit and seed production.
- **Butterflies (Lepidoptera):** Important pollinators, especially for flowering plants.
- **Hoverflies (Syrphidae):** Both pollinators and predators of aphids.

2. Predators:
- **Ladybugs (Coccinellidae):** Feed on aphids, mites, and other soft-bodied insects.
- **Lacewings (Chrysopidae):** Larvae consume aphids, caterpillars, and thrips.
- Spiders (Araneae): Generalist predators that capture a variety of garden pests.

3. Parasitizers:
- **Parasitic Wasps (Ichneumonidae, Braconidae):** Lay eggs inside or on pest insects, eventually killing the host.

- **Tachinid Flies (Tachinidae):** Parasitize caterpillars and beetles.

4. Decomposers:
   - **Beetles (Coleoptera):** Some species, like ground beetles, help break down organic matter and prey on pests.
   - **Ants (Formicidae):** Aid in decomposing organic material and preying on small insects.

Creating a Habitat for Beneficial Insects

1. Diverse Plantings:
   - **Plant Diversity:** Grow a variety of plants to provide continuous blooms throughout the growing season, ensuring a constant food source for pollinators and other beneficial insects.
   - **Native Plants:** Incorporate native plants, as they are well-adapted to local conditions and more attractive to native beneficial insects.

2. Flowering Plants:

- **Pollinator-Friendly Flowers:** Include flowers that are rich in nectar and pollen. Examples include sunflowers, daisies, coneflowers, and marigolds.
- **Herbs:** Many herbs, such as dill, fennel, and cilantro, produce flowers that attract beneficial insects.

3. Habitat Structures:

- **Insect Hotels:** Create insect hotels or shelters to provide nesting sites and overwintering habitats for beneficial insects.
- **Mulch and Ground Covers:** Use organic mulch and ground covers to provide habitat for ground-dwelling predators and decomposers.

4. Water Sources:

- Shallow Water: Provide shallow water sources, such as birdbaths with pebbles or shallow dishes, to offer drinking water for beneficial insects.

5. Avoiding Pesticides:

- Reduce Chemical Use: Minimize or eliminate the use of chemical pesticides, which can harm beneficial insects. Opt for organic pest control methods and integrated pest management (IPM) practices.

Attracting Specific Beneficial Insects

1. Attracting Pollinators:

- **Bees:** Plant a mix of flowers that bloom at different times. Provide nesting sites such as hollow stems, bare ground, and bee houses.
- **Butterflies:** Grow nectar-rich plants and host plants for caterpillars, like milkweed for monarchs.

2. Attracting Predators:

- **Ladybugs**: Plant flowering plants like yarrow, dill, and fennel. Provide a diverse garden that supports a variety of prey.

- **Lacewings:** Grow plants with small flowers, such as dill, fennel, and cosmos, which attract adult lacewings.

3. Attracting Parasitizers:
- Parasitic Wasps: Plant flowering herbs and members of the carrot family (Apiaceae), such as parsley and Queen Anne's lace, which provide nectar for adult wasps.
- **Tachinid Flies:** Grow plants like sweet alyssum and dill to attract these parasitic flies.

Maintaining Beneficial Insect Populations

1. Continuous Bloom:
- Seasonal Planting: Ensure a succession of blooming plants to provide a consistent food source throughout the growing season.
- **Perennial Plants:** Incorporate perennials that bloom in different seasons to maintain a year-round habitat.

2. Habitat Protection:

- Minimize Disturbance: Limit soil disturbance and tilling to protect ground-dwelling insects and their habitats.

- Leaf Litter: Leave leaf litter and plant debris in place during the winter to provide overwintering sites for beneficial insects.

3. Integrated Pest Management (IPM):

- Monitoring: Regularly monitor pest populations and beneficial insect activity to make informed pest management decisions.

- **Biological Controls**: Use biological control methods, such as releasing beneficial insects, to manage pest populations naturally.

4. Education and Awareness:

- Community Involvement: Educate yourself and your community about the importance of beneficial insects and how to support them in the garden.

- Garden Planning: Plan your garden with beneficial insects in mind, considering their habitat needs and food sources.

Conclusion

Encouraging beneficial insects in your permaculture garden is essential for creating a healthy, balanced, and sustainable ecosystem. By understanding the types of beneficial insects and their roles, creating diverse and supportive habitats, and adopting practices that minimize harm, you can enhance the presence and effectiveness of these vital allies. Beneficial insects contribute to natural pest control, pollination, soil health, and overall garden productivity, making them indispensable partners in any successful permaculture system. Through thoughtful planning and ongoing care, your garden can thrive with the help of beneficial insects, leading to a more resilient and bountiful environment.

Organic Pest Management

Introduction to Organic Pest Management

Organic pest management is a holistic approach that focuses on maintaining the health of the garden ecosystem to prevent and control pest problems without relying on synthetic chemicals. This approach promotes biodiversity, soil health, and the use of natural predators and organic treatments. The goal is to create a balanced ecosystem where pests are kept in check naturally, reducing the need for interventions and fostering a resilient and sustainable garden.

Principles of Organic Pest Management

1. Prevention: The foundation of organic pest management is preventing pest problems before they arise. This includes choosing the right plants, maintaining healthy soil, and fostering a diverse ecosystem.

2. Observation: Regular monitoring of your garden helps detect pest problems early, allowing for timely and effective interventions.

3. Intervention: When pest problems do occur, organic pest management focuses on the least disruptive solutions, prioritizing biological, mechanical, and cultural methods over chemical treatments.

Key Strategies for Organic Pest Management

1. Healthy Soil:
   - **Soil Health:** Healthy soil is the cornerstone of a healthy garden. Use compost, organic matter, and natural fertilizers to build fertile, well-draining soil.

- **Soil Testing**: Regular soil testing helps you understand the nutrient needs and pH balance of your soil, allowing for appropriate amendments.

2. Plant Selection:
- **Diverse Planting:** Plant a variety of species to create a resilient ecosystem that can withstand pest pressures. Diverse plantings also support beneficial insects and natural predators.
- **Resistant Varieties:** Choose plant varieties that are resistant to common pests and diseases in your area.

3. Companion Planting:
- **Beneficial Relationships**: Use companion planting to enhance pest control. Some plants can repel pests, attract beneficial insects, or provide a physical barrier to pests.
- **Examples**: Marigolds can repel nematodes, while basil can deter aphids and mosquitoes. Planting nasturtiums can

attract aphids away from more valuable crops.

4. Crop Rotation:
   - **Disrupting Pest Cycles:** Rotate crops each season to prevent the build-up of pests and diseases associated with specific plants. This disrupts the life cycles of pests and reduces their populations.
   - **Rotation Plans**: Develop a crop rotation plan that moves plant families to different locations in the garden each year.

5. Physical Barriers and Traps:
   - **Row Covers:** Use floating row covers to protect plants from insects and other pests while allowing light and water to penetrate.
   - **Traps:** Deploy traps such as sticky traps, pheromone traps, or physical traps to capture and reduce pest populations.

6. Biological Controls:
   - **Beneficial Insects**: Introduce or encourage beneficial insects like ladybugs,

lacewings, and parasitic wasps that prey on pest insects.

- **Predatory Nematodes:** Use predatory nematodes to target soil-dwelling pests such as grubs and root maggots.

- **Microbial Insecticides:** Utilize microbial insecticides like Bacillus thuringiensis (Bt), which targets specific pests without harming beneficial insects.

7. Natural Predators:

- **Attracting Predators**: Create habitats that attract natural predators such as birds, bats, and amphibians. Planting trees and shrubs, installing birdhouses, and maintaining water sources can help.

- **Predator Diversity:** Encourage a variety of predators to ensure that no single pest species dominates.

8. Organic Pesticides:

- **Botanical Insecticides**: Use plant-based insecticides such as neem oil, pyrethrin, and insecticidal soaps. These products are derived from natural sources and are less harmful to non-target organisms.

- **Mineral-based Pesticides:** Use mineral-based products like diatomaceous earth and sulfur to control pests. These substances are effective but must be used carefully to avoid harming beneficial insects.

Specific Organic Pest Management Techniques

1. Neem Oil:

- **Usage:** Neem oil is an effective organic pesticide that controls a wide range of pests, including aphids, mites, and whiteflies. It disrupts the life cycle of pests and acts as a repellent.

- **Application:** Apply neem oil as a foliar spray, ensuring thorough coverage of leaves and stems. Reapply as needed, particularly after rain.

2. Insecticidal Soap:

- **Usage:** Insecticidal soap is a contact insecticide that controls soft-bodied insects like aphids, mealybugs, and spider mites. It disrupts the cell membranes of insects, leading to their death.

- **Application:** Spray insecticidal soap directly on affected plants, targeting the undersides of leaves where pests often hide. Repeat applications may be necessary.

3. Diatomaceous Earth:

- **Usage**: Diatomaceous earth is a natural powder made from fossilized algae. It damages the exoskeletons of insects, causing them to dehydrate and die.

- **Application:** Dust diatomaceous earth around the base of plants and on soil surfaces where pests are present. Reapply after rain or watering.

4. Companion Planting Examples:

- **Marigolds and Tomatoes:** Marigolds can repel nematodes and attract beneficial insects that prey on tomato pests.

- **Basil and Peppers:** Basil can deter aphids, spider mites, and mosquitoes, benefiting nearby pepper plants.

- **Nasturtiums and Cucumbers:** Nasturtiums attract aphids away from cucumbers and can also repel cucumber beetles.

5. Beneficial Insect Habitat:

- **Flowering Plants**: Plant flowers like yarrow, dill, and fennel to attract beneficial insects. These plants provide nectar and pollen, supporting predator and parasitoid populations.

- **Insect Hotels:** Create insect hotels or shelters for solitary bees, lacewings, and other beneficial insects. Use materials like bamboo, straw, and wood with drilled holes.

Monitoring and Record-Keeping

1. Regular Scouting:

- **Observation:** Regularly inspect your garden for signs of pest activity, damage, and beneficial insect presence. Early detection allows for timely interventions.

- **Identification:** Accurately identify pests and beneficial insects to choose the most effective management strategies.

2. Record-Keeping:

- **Pest Logs**: Maintain records of pest sightings, plant health, and interventions used. This helps track patterns and effectiveness over time.

- **Garden Journal:** Keep a garden journal to document crop rotations, companion planting success, and soil health improvements.

Integrated Pest Management (IPM)

1. IPM Approach:

- **Combination of Methods:** Integrate multiple pest management strategies to create a comprehensive and resilient approach. Combine cultural, mechanical, biological, and organic methods.

- **Thresholds:** Establish pest thresholds, which are levels of pest activity that trigger interventions. This prevents unnecessary treatments and preserves beneficial insect populations.

2. Education and Community Involvement:

- **Learning**: Continuously educate yourself about organic pest management techniques and best practices. Stay informed about new research and innovations.

- **Sharing Knowledge:** Share your knowledge with your gardening community. Collaborative efforts can enhance the overall health and productivity of local gardens.

Conclusion

Organic pest management is an essential component of sustainable gardening and permaculture. By focusing on prevention, observation, and the use of natural methods, gardeners can create healthy, balanced ecosystems that minimize pest problems. Healthy soil, diverse plantings, companion planting, crop rotation, physical barriers, biological controls, and organic pesticides all play a role in this holistic approach. Regular monitoring, record-keeping, and an integrated pest management strategy ensure that interventions are effective and minimally disruptive. Through education and community involvement, gardeners can continue to improve their organic pest management practices, contributing to a resilient and productive garden environment.

2. Crop Rotation and Diversity

Introduction to Crop Rotation and Diversity

Crop rotation and plant diversity are foundational practices in sustainable gardening and permaculture. These techniques contribute to soil health, pest and disease management, nutrient balance, and ecosystem resilience. By strategically rotating crops and incorporating a diverse range of plants, gardeners can enhance the productivity and sustainability of their gardens.

The Concept of Crop Rotation

1. Definition:
Crop rotation is the practice of growing different types of crops in the same area in sequential seasons. This planned sequence aims to reduce soil nutrient depletion, manage pests and diseases, and improve soil structure and fertility.

2. Historical Background:
The concept of crop rotation dates back to ancient agricultural practices. Early civilizations recognized the benefits of rotating crops to maintain soil fertility and reduce pest pressures. The three-field system of medieval Europe, where fields were divided into sections and rotated between different crops and fallow periods, is a notable historical example.

3. Benefits of Crop Rotation:

- **Soil Fertility**: Different crops have varying nutrient requirements and contribute differently to soil health. Legumes, for example, fix nitrogen in the soil, enriching it for subsequent crops.
- **Pest and Disease Management:** Rotating crops interrupts the life cycles of pests and diseases associated with specific plants. This reduces the build-up of pest populations and pathogen loads in the soil.
- **Weed Suppression:** Crop rotation can help manage weed populations by disrupting their growth cycles and reducing their chances of establishing dominance.
- **Soil Structure:** Different root structures and growth habits of crops improve soil structure, promote aeration, and enhance water infiltration.

Principles of Effective Crop Rotation

1. **Grouping Plants by Family:**
   - **Botanical Families:** Rotate crops based on their botanical families to avoid planting closely related plants in the same area consecutively. This reduces the risk of pest and disease carryover.
   - **Examples:** Group crops into families such as Solanaceae (tomatoes, peppers), Brassicaceae (cabbage, broccoli), Fabaceae (beans, peas), and Asteraceae (lettuce, sunflowers).

2. **Rotating by Nutrient Use:**
   - Heavy Feeders, Light Feeders, and Soil Builders: Rotate crops based on their nutrient requirements. Heavy feeders (e.g., corn, tomatoes) should be followed by light feeders (e.g., root vegetables) and then soil builders (e.g., legumes).
   - **Nutrient Management:** This rotation helps prevent nutrient depletion and ensures a balanced nutrient profile in the soil.

3. Incorporating Cover Crops:

- **Benefits of Cover Crops:** Include cover crops in your rotation to improve soil health, prevent erosion, suppress weeds, and add organic matter. Examples include clover, rye, and buckwheat.

- **Green Manure:** Some cover crops, like legumes, can be turned into the soil as green manure to enhance soil fertility.

4. Length of Rotation Cycles:

- **Multi-Year Cycles:** Plan crop rotations over several years. A typical cycle might range from three to five years, depending on the crops and garden size.

- **Record Keeping**: Keep detailed records of your crop rotations to track plant placements, monitor soil health, and plan future rotations.

The Role of Plant Diversity

1. Definition and Importance:

Plant diversity refers to growing a variety of plant species in your garden. This diversity supports ecosystem health, promotes beneficial interactions, and enhances garden resilience.

2. Benefits of Plant Diversity:

- **Pest Management:** Diverse plantings attract beneficial insects and natural predators, reducing pest populations and minimizing damage.
- **Disease Resistance:** A variety of plants can hinder the spread of diseases by reducing the availability of hosts for pathogens.
- **Nutrient Cycling:** Different plants contribute to and draw from the soil in unique ways, promoting balanced nutrient cycling and soil health.
- **Microclimates:** Diverse plant structures create microclimates within the garden, moderating temperature and humidity and improving plant growth conditions.
- **Pollinator Support**: A variety of flowering plants ensures continuous food sources for

pollinators, enhancing pollination and fruit set.

Implementing Plant Diversity

1. Polycultures:
   - **Definition:** Polyculture involves growing multiple crop species in the same area, mimicking natural ecosystems.
   - **Benefits:** Polycultures enhance biodiversity, improve soil health, and provide habitat for beneficial insects and organisms.

2. Intercropping:
   - **Definition:** Intercropping is the practice of growing two or more crops in proximity to each other.
   - **Examples:** Planting beans with corn (the "Three Sisters" method) or growing carrots with onions to deter pests and maximize space.
   - **Advantages:** Intercropping can lead to better resource utilization, pest suppression, and increased overall yield.

3. Companion Planting:

- **Beneficial Relationships:** Pair plants that benefit each other. For instance, marigolds repel nematodes, while basil deters aphids and enhances the growth of tomatoes.

- **Guilds:** Create plant guilds, which are groups of plants that support each other through nutrient sharing, pest control, and other beneficial interactions.

4. Succession Planting:

- **Continuous Harvest:** Plan for a continuous harvest by planting crops in succession. This maximizes space and ensures a steady supply of produce.

- **Examples**: Follow early-harvested crops like radishes with later-season crops like beans or squash.

5. Habitat Diversity:

- **Wildlife Habitat:** Incorporate native plants, flowering perennials, and shrubs to create habitats for wildlife and beneficial insects.

- **Microhabitats:** Create different microhabitats within the garden by varying plant heights, densities, and structures.

Case Studies and Examples

1. Traditional Practices:

- **Indigenous Agriculture:** Many indigenous agricultural practices, such as the "Three Sisters" method used by Native American tribes, exemplify effective crop rotation and plant diversity strategies.

- **Asian Rice-** Integrated rice and fish farming systems in Asia demonstrate how polyculture and biodiversity enhance productivity and sustainability.

2. Modern Applications:

- **Organic Farms:** Successful organic farms often utilize crop rotation and diversity to maintain soil health and manage pests without synthetic chemicals.

- **Permaculture Gardens:** Permaculture design principles emphasize diversity and rotation to create self-sustaining ecosystems.

Practical Steps for Gardeners

1. Planning and Design:
- **Garden Layout**: Design your garden to facilitate crop rotation and diversity. Create beds or zones that can be easily rotated each season.
- **Plant Lists:** Develop plant lists grouped by botanical family and nutrient requirements to assist with rotation planning.

2. Soil Management:
- **Compost and Mulch**: Regularly add compost and organic mulch to maintain soil fertility and structure.
- **Soil Testing:** Conduct soil tests periodically to monitor nutrient levels and pH, adjusting your rotation and amendment plans accordingly.

3. Record Keeping:

- Garden Journal: Maintain a garden journal to document plant placements, rotation schedules, soil conditions, and pest and disease observations.
- **Maps and Diagrams:** Use maps and diagrams to visualize crop rotations and plan for future seasons.

4. Community Involvement:

- **Knowledge Sharing**: Share your experiences and learn from other gardeners and farmers. Community gardens and local agricultural organizations can be valuable resources.
- **Workshops and Education:** Participate in workshops and educational programs on organic gardening, permaculture, and sustainable agriculture.

Conclusion

Crop rotation and plant diversity are essential strategies for sustainable gardening. By understanding and implementing these practices, gardeners can enhance soil health, manage pests and diseases, and promote a balanced and resilient ecosystem. Whether through traditional methods or modern applications, the principles of rotation and diversity contribute to the long-term productivity and sustainability of gardens and farms. By planning carefully, maintaining healthy soil, and fostering a diverse range of plants, gardeners can create thriving, self-sustaining systems that benefit both the environment and their harvests.

Preventing Soil Depletion

Introduction to Soil Depletion

Soil depletion is the process by which soil loses its nutrients, organic matter, and overall fertility due to various factors such as continuous cropping, overgrazing, deforestation, and improper agricultural practices. This degradation of soil health compromises plant growth, reduces crop yields, and affects the entire ecosystem. Preventing soil depletion is crucial for sustainable gardening, agriculture, and long-term environmental health.

Causes of Soil Depletion

1. Monocropping:
   - **Definition**: The practice of growing the same crop in the same area year after year.
   - **Impact**: Leads to the depletion of specific nutrients that the crop consistently uses, resulting in nutrient imbalances and soil exhaustion.

2. Overgrazing:
   - **Definition:** When livestock graze excessively on a piece of land, not allowing plants enough time to recover.

- **Impact:** Reduces plant cover, leading to soil erosion, compaction, and loss of soil organic matter.

3. Intensive Tillage:
- **Definition:** Frequent and deep plowing of soil.
- **Impact:** Disrupts soil structure, increases erosion, and depletes organic matter, reducing soil fertility.

4. Chemical Fertilizers:
- **Overuse**: Excessive use of synthetic fertilizers can lead to nutrient imbalances, soil acidification, and reduction of beneficial soil microorganisms.
- **Dependency**: Long-term reliance on chemical fertilizers diminishes soil's natural fertility and resilience.

5. Soil Erosion:
- **Causes:** Wind and water erosion remove the topsoil, which is the most fertile

layer containing the highest concentration of organic matter and nutrients.

   - **Consequences**: Loss of topsoil significantly reduces soil productivity and its capacity to support plant growth.

6. Depletion of Soil Organic Matter:
   - **Importance**: Organic matter is crucial for maintaining soil structure, water retention, and nutrient cycling.

   - **Loss Factors:** Practices like burning crop residues, leaving soil bare, and over-tillage contribute to the depletion of soil organic matter.

Strategies to Prevent Soil Depletion

1. Crop Rotation:
   - **Diverse Planting**: Rotate different crops to balance nutrient use and prevent the build-up of pests and diseases.

- **Nutrient Cycling**: Include legumes in rotation to fix nitrogen in the soil, improving fertility for subsequent crops.

2. Cover Crops:
- **Soil Protection**: Plants cover crops during off-seasons to protect soil from erosion, improve organic matter, and enhance soil structure.
- **Examples:** Rye, clover, and vetch are effective cover crops that can be turned into the soil as green manure.

3. Reduced Tillage:
- **Conservation Tillage**: Minimize tillage to maintain soil structure, reduce erosion, and preserve soil organic matter.
- **No-Till Farming:** Adopt no-till or low-till practices where feasible to enhance soil health and organic matter retention.

4. Organic Amendments:
- **Compost:** Regularly add compost to enrich soil with organic matter and nutrients.

- **Manure:** Use well-aged manure to improve soil fertility and microbial activity.

5. Mulching:
- **Soil Cover:** Apply organic mulches like straw, leaves, or wood chips to protect soil from erosion, retain moisture, and add organic matter as they decompose.
- **Weed Suppression**: Mulching also helps suppress weed growth, reducing competition for nutrients and water.

6. Soil Testing and Amendments:
- **Regular Testing:** Conduct soil tests to monitor nutrient levels, pH, and organic matter content.
- **Balanced Amendments**: Apply lime, gypsum, or other soil amendments based on test results to correct nutrient imbalances and improve soil health.

7. Composting:

- **Waste Recycling**: Compost garden waste, kitchen scraps, and other organic materials to create nutrient-rich compost.
  - **Soil Enrichment**: Regularly incorporate compost into the soil to boost organic matter and fertility.

8. **Polyculture and Biodiversity:**
   - **Diverse Planting**: Grow a variety of plants together to enhance biodiversity and improve soil health.
   - **Companion Planting**: Use companion planting to support soil health and pest management.

9. **Soil Conservation Techniques:**
   - **Terracing:** Implement terracing on slopes to reduce erosion and retain soil.
   - **Contour Plowing:** Plow along contour lines to slow water runoff and minimize erosion.

10. **Agroforestry:**

- **Integration**: Integrate trees and shrubs into agricultural systems to improve soil stability, enhance biodiversity, and increase organic matter.
- **Benefits:** Trees provide shade, reduce wind erosion, and contribute to nutrient cycling through leaf litter.

11. Organic Fertilizers:
- **Natural Nutrients:** Use organic fertilizers like bone meal, blood meal, and fish emulsion to provide slow-release nutrients.
- **Soil Health**: Organic fertilizers improve soil structure and microbial activity.

12. Managed Grazing:
- **Rotational Grazing:** Rotate livestock to prevent overgrazing and allow vegetation recovery.

- **Benefits:** Managed grazing improves soil structure, increases organic matter, and enhances nutrient cycling.

13. Water Management:
- **Efficient Irrigation:** Use efficient irrigation systems like drip irrigation to reduce water stress and prevent soil erosion.
- **Rainwater Harvesting:** Collect and use rainwater to minimize reliance on groundwater and reduce runoff.

14. Soil Health Monitoring:
- **Regular Assessments**: Monitor soil health indicators such as organic matter content, microbial activity, and soil structure.
- **Adaptive Management:** Adjust practices based on monitoring results to continuously improve soil health.

Implementing Soil Conservation Practices

1. Planning:

- **Soil Mapping:** Conduct a detailed assessment of soil types, erosion risks, and nutrient levels in your garden.
- **Site-Specific Strategies:** Develop soil conservation plans tailored to the specific conditions and needs of your garden.

2. Community Involvement:
- **Knowledge Sharing:** Engage with local gardening communities to share knowledge and experiences related to soil conservation.
- **Collaborative Projects**: Participate in community projects aimed at promoting sustainable gardening practices.

3. Education and Training:
- Workshops and Seminars: Attend workshops and seminars on soil health, conservation techniques, and sustainable agriculture.
- **Online Resources**: Utilize online resources, courses, and forums to stay

informed about best practices and new developments in soil conservation.

Conclusion

Preventing soil depletion is a fundamental aspect of sustainable gardening and agriculture. By adopting practices such as crop rotation, cover cropping, reduced tillage, organic amendments, and mulching, gardeners can maintain and enhance soil health. Regular soil testing, the use of organic fertilizers, and efficient water management further contribute to preventing soil depletion. Emphasizing biodiversity through polycultures, agroforestry, and managed grazing also supports resilient and productive soils. By implementing these strategies and continuously monitoring soil health, gardeners can ensure long-term soil fertility, productivity, and environmental sustainability.

Enhancing Biodiversity

Introduction to Biodiversity in Permaculture

Biodiversity refers to the variety and variability of life forms within a given ecosystem, including the diversity of species, genes, and ecosystems. In permaculture gardening, enhancing biodiversity is a foundational principle. It promotes resilience, sustainability, and productivity by mimicking natural ecosystems. Biodiverse systems are better equipped to withstand environmental stresses, pests, and diseases, and they provide a wider range of ecological services.

Importance of Biodiversity

1. Ecosystem Stability:

- **Resilience:** Diverse ecosystems are more resilient to disturbances such as pests, diseases, and extreme weather events.
- **Balance:** High biodiversity helps maintain ecological balance by promoting natural pest control and nutrient cycling.

2. Soil Health:
- **Nutrient Cycling:** Diverse plant species contribute to a more complex and efficient nutrient cycle, enriching soil fertility.
- **Organic Matter**: A variety of plants contribute different types of organic matter, enhancing soil structure and health.

3. Pest and Disease Control:
- **Natural Predators**: A diverse ecosystem supports a range of natural predators and beneficial insects that help control pest populations.
- **Disease Resistance**: Genetic diversity within plant species reduces the spread and impact of diseases.

4. Pollination:

- **Pollinator Diversity**: Diverse plant species attract a wide range of pollinators, improving pollination rates and fruit set.

- **Habitat Provision**: A variety of flowering plants provide habitat and food for pollinators throughout the growing season.

5. Water Management:

- **Water Retention**: Diverse plant roots enhance soil structure and water retention, reducing runoff and erosion.

- **Hydrological Cycles**: Biodiverse systems support balanced hydrological cycles, promoting sustainable water use.

Strategies to Enhance Biodiversity

1. Polyculture:

- **Definition**: Growing multiple crop species together in the same space.
- **Benefits**: Enhances ecological interactions, reduces pest outbreaks, and improves resource use efficiency.
- **Implementation:** Combine complementary plants that support each other through nutrient exchange, shade provision, and pest control.

2. Companion Planting:

- **Definition**: Planting different species in close proximity for mutual benefit.
- **Examples:**
    - **Three Sisters**: Corn, beans, and squash. Corn provides support for beans, beans fix nitrogen for the soil, and squash covers the ground to suppress weeds.
    - Tomatoes and Basil: Basil repels pests and enhances the flavor of tomatoes.
- **Benefits:** Improves plant health, boosts yields, and deters pests.

3. Plant Guilds:

- **Definition:** Groupings of plants that work together synergistically.
- **Components:** Typically include a central plant (often a tree), support species (nitrogen fixers, nutrient accumulators), and ground covers.
- **Examples**: Apple tree guild with comfrey, clover, and daffodils.
- **Benefits:** Enhances soil health, promotes biodiversity, and increases resilience.

4. Native and Adaptable Species:

- **Native Plants:** Species that naturally occur in a specific region and are well-adapted to local conditions.
- **Adaptable Species**: Non-native species that can thrive in local conditions without becoming invasive.
- **Benefits**: Native and adaptable species support local wildlife, require less maintenance, and are more resilient to local climate and soil conditions.

5. Perennials and Annuals:

- **Perennials**: Plants that live for more than two years, providing long-term stability and continuous cover.
- **Annuals:** Plants that complete their life cycle in one growing season, offering flexibility and diversity in crop planning.
- **Benefits:** Combining perennials and annuals ensures continuous cover, diverse root structures, and varied ecological niches.

6. Creating Habitats:

- **Wildlife Habitats**: Incorporate features such as ponds, hedgerows, and rock piles to provide habitat for beneficial wildlife.
- **Insect Hotels**: Construct shelters for beneficial insects like bees, ladybugs, and predatory beetles.
- **Birdhouses:** Install birdhouses to attract insect-eating birds, contributing to natural pest control.

7. Integrating Animals:

- **Poultry:** Chickens, ducks, and other poultry can help control pests, fertilize the soil, and provide eggs and meat.
- **Grazing Animals**: Goats, sheep, and other grazing animals can manage weeds, improve soil fertility, and provide manure.
- **Bees:** Beekeeping enhances pollination and provides honey.

8. Succession Planting:

- **Definition**: Planting crops in succession to ensure continuous harvest and soil cover.
- **Examples:** Planting early spring crops followed by summer and fall crops.
- **Benefits**: Maximizes productivity, reduces bare soil periods, and maintains continuous biodiversity.

9. Seasonal Planting:

- **Cool-Season Crops**: Planting crops that thrive in cooler temperatures in early spring and fall.

- **Warm-Season Crops**: Planting crops that require warmer temperatures in late spring and summer.
- **Benefits:** Extends the growing season, optimizes resource use, and increases plant diversity.

10. Edible Landscapes:
- **Definition**: Integrating food-producing plants into ornamental landscapes.
- **Examples:** Fruit trees, berry bushes, and vegetable beds mixed with ornamental plants.
- **Benefits**: Enhances biodiversity, aesthetic appeal, and food production.

11. Encouraging Beneficial Insects:
- **Flowering Plants:** Plant a variety of flowering plants to attract pollinators and beneficial insects.
- **Herbs**: Include herbs like dill, fennel, and coriander that attract predatory insects.

- **Habitat Creation**: Provide habitats such as insect hotels and undisturbed areas for beneficial insects.

12. Rotational Grazing:
- **Definition:** Moving livestock between pastures to allow vegetation recovery.
- **Benefits:** Enhances pasture biodiversity, prevents overgrazing, and improves soil health.

13. Soil Health:
- **Organic Matter**: Regularly add compost and organic matter to improve soil structure and support a diverse soil microbiome.
- **Cover Crops:** Use cover crops to protect and enhance soil biodiversity during off-seasons.
- **Reduced Tillage**: Minimize soil disturbance to preserve soil structure and biodiversity.

14. Community Involvement:

- **Knowledge Sharing:** Engage with local gardening communities to share knowledge and experiences related to enhancing biodiversity.

- **Collaborative Projects:** Participate in community projects aimed at promoting biodiversity and sustainable gardening practices.

Conclusion

Enhancing biodiversity in permaculture gardening is essential for creating resilient, sustainable, and productive ecosystems. By adopting practices such as polyculture, companion planting, plant guilds, and integrating native and adaptable species, gardeners can promote a rich and diverse ecosystem. Incorporating perennials, creating habitats, and encouraging beneficial insects further enhance biodiversity and ecological balance. These strategies, combined with careful soil and water management, ensure long-term soil health, pest control, and sustainable productivity. Embracing biodiversity not only supports the health of the garden but also contributes to the broader environmental sustainability and resilience.

CHAPTER 7

Tools and Resources

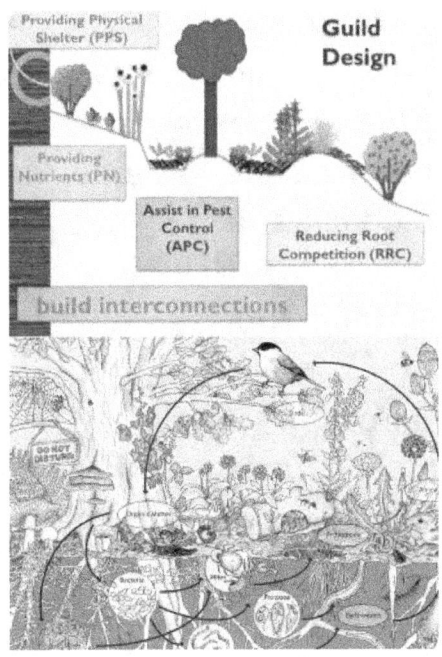

Introduction to Permaculture Tools and Resources

Permaculture gardening requires a blend of traditional gardening tools, specialized equipment, and various resources that facilitate sustainable practices. These tools and resources help in planning, designing, implementing, and maintaining a permaculture garden. From basic hand tools to advanced soil testing kits, and from educational resources to community networks, the right tools and resources can make permaculture gardening more efficient, productive, and enjoyable.

Essential Hand Tools

1. Garden Fork:
 - **Uses:** Turning and aerating soil, breaking up clods, and incorporating organic matter.
 - **Benefits**: Essential for soil preparation and maintaining soil structure.

2. Spade and Shovel:
 - **Uses:** Digging, planting, moving soil, and compost.

- **Benefits**: Versatile tools for a variety of garden tasks, from planting trees to turning compost.

3. Hand Trowel:
- **Uses:** Planting seedlings, transplanting, and digging small holes.
- **Benefits**: Ideal for detailed work in small spaces and container gardens.

4. Pruners and Shears:
- **Uses:** Pruning shrubs, trees, and perennials, harvesting fruits and vegetables.
- **Benefits**: Ensures healthy plant growth and easy harvesting.

5. Rake:
- **Uses**: Smoothing soil, spreading mulch, collecting leaves and debris.
- **Benefits:** Maintains tidy garden beds and pathways, helps incorporate organic matter.

6. Hoe:
   - **Uses**: Weeding, cultivating soil, creating planting rows.
   - **Benefits**: Reduces weed competition and prepares soil for planting.

7. Wheelbarrow or Garden Cart:
   - **Uses:** Transporting soil, compost, plants, and tools around the garden.
   - **Benefits:** Saves time and effort, particularly in larger gardens.

8. Watering Can and Hose:
   - **Uses:** Irrigating plants, applying liquid fertilizers.
   - **Benefits:** Ensures efficient water delivery, particularly in areas without automated irrigation.

9. Mulch Fork:
   - **Uses**: Spreading mulch, turning compost.
   - **Benefits:** Eases the distribution of bulky organic materials.

Advanced Tools and Equipment

1. Soil Testing Kit:
- **Uses:** Analyzing soil pH, nutrient levels, and soil composition.
- **Benefits:** Informs soil amendment and fertilization strategies, ensuring optimal soil health.

2. Drip Irrigation System:
- **Uses:** Efficiently delivering water directly to plant roots.
- **Benefits:** Conserves water, reduces evaporation, and minimizes weed growth.

3. Rainwater Harvesting System:
- **Uses:** Collecting and storing rainwater for garden use.
- **Benefits**: Provides a sustainable water source, reduces reliance on municipal water, and lowers water bills.

4. Compost Bin or Tumbler:
   - **Uses:** Creating compost from garden waste, kitchen scraps, and other organic materials.
   - **Benefits:** Produces nutrient-rich compost, reduces waste, and improves soil health.

5. Raised Beds:
   - **Uses:** Growing plants in controlled environments with improved soil quality and drainage.
   - **Benefits:** Enhances plant growth, eases access for planting and harvesting, and reduces soil compaction.

6. Greenhouse or Polytunnel:
   - **Uses:** Extending the growing season, protecting plants from extreme weather.
   - **Benefits:** Allows for year-round gardening, increases plant variety, and improves yields.

7. Solar Dehydrator:

- **Uses:** Drying herbs, fruits, and vegetables using solar energy.

- **Benefits:** Preserves produce, reduces food waste, and utilizes renewable energy.

Resources for Learning and Support

1. Books and Publications:
- **Examples:**
  - "Permaculture: A Designer's Manual" by Bill Mollison
  - **"Gaia's Garden**: A Guide to Home-Scale Permaculture" by Toby Hemenway
  - "The Resilient Farm and Homestead" by Ben Falk

- **Benefits:** Comprehensive guides on permaculture principles, design, and practices.

2. Online Courses and Webinars:
  - Examples
   - Permaculture Design Certificate (PDC) courses offered by various institutions and online platforms.
   - Webinars hosted by permaculture experts and organizations.
   - **Benefits:** Flexible learning options, access to expert knowledge, and opportunities for interactive learning.

3. Websites and Blogs:
  - Examples:
   - Permaculture Institute (permaculture.org)
   - Permaculture Research Institute (permaculturenews.org)
   - Rich Soil (richsoil.com)
   - **Benefits:** Up-to-date information, practical tips, and community forums for discussion and support.

4. YouTube Channels:
  - **Examples:**
    - **"Living Web Farms"** - Demonstrations of permaculture techniques and practices.
    - **"Happen Films"** - Documentaries and stories about permaculture projects worldwide.
    - **"Morag Gamble:** Our Permaculture Life" - Educational videos on permaculture gardening and living.
    - **Benefits**: Visual learning, practical demonstrations, and inspirational content.

5. Local Permaculture Groups and Networks:
  - **Examples**: Local permaculture clubs, gardening cooperatives, and permaculture networks.
  - **Benefits:** Opportunities for networking, knowledge exchange, and collaboration on projects.

6. Workshops and Events:
- **Examples**: Hands-on workshops, permaculture conferences, and community garden events.
- **Benefits:** Practical experience, skill-building, and connecting with like-minded individuals.

7. Extension Services and Local Agricultural Offices:
- **Uses**: Accessing expert advice, soil testing services, and educational resources.
- **Benefits:** Tailored support for local conditions, access to research-based information.

Sustainable Practices and Resources

1. Seed Saving:
- **Practice**: Collecting and storing seeds from plants for future planting.
- **Benefits:** Preserves genetic diversity, reduces seed costs, and ensures seed availability.

2. Organic Mulches:
- **Examples:** Straw, wood chips, leaves, and grass clippings.
- **Benefits:** Suppresses weeds, retains soil moisture, and adds organic matter to the soil.

3. Natural Pest Control:
- **Examples**: Beneficial insects, biological controls, and natural repellents.
- **Benefits:** Reduces chemical use, promotes ecological balance, and protects beneficial organisms.

4. Renewable Energy Resources:
- **Examples:** Solar panels for powering garden equipment, solar water heaters.
- **Benefits:** Reduces carbon footprint, lowers energy costs, and promotes sustainable living.

5. Composting Toilets:
   - Uses: Converting human waste into compost.
   - **Benefits:** Reduces water use, recycles nutrients, and provides a sustainable waste management solution.

Conclusion

Equipping yourself with the right tools and resources is essential for successful permaculture gardening. From basic hand tools to advanced equipment, and from educational resources to community networks, these tools and resources facilitate sustainable practices, enhance productivity, and promote ecological balance. By leveraging a combination of practical tools, learning materials, and support networks, permaculture gardeners can create resilient, productive, and sustainable ecosystems.

1. Essential Tools for Permaculture Gardening

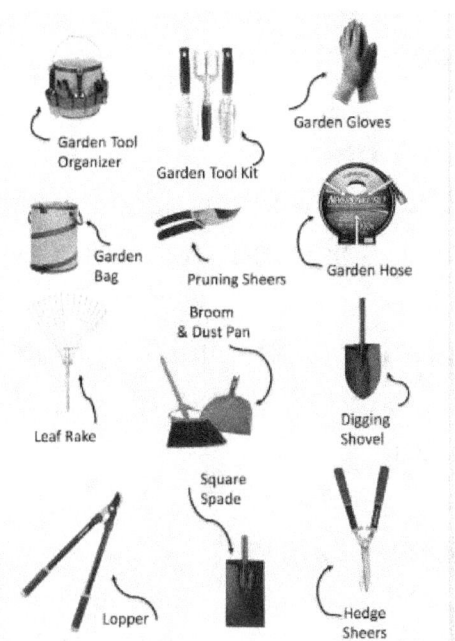

Permaculture gardening is a holistic approach to land management that focuses on creating sustainable and self-sufficient ecosystems. To achieve this, gardeners use a variety of tools that help them design, implement, and maintain their gardens efficiently. This guide will cover essential tools for soil preparation, planting, maintenance, and sustainable practices in permaculture gardening.

Soil Preparation Tools

1. Garden Fork:
   - **Uses:** Aerating soil, breaking up compacted soil, and incorporating organic matter.
   - **Benefits:** Essential for improving soil structure and ensuring healthy root development.

2. Spade and Shovel:
  - **Spade:**
     - **Uses:** Digging planting holes, edging garden beds, and cutting through tough soil.
     - **Benefits:** Versatile tool for precise digging and soil preparation.
  - Shovel:
     - **Uses:** Moving soil, compost, and mulch.

- **Benefits**: Efficient for handling large volumes of material.

3. Broadfork:
- **Uses:** Loosening soil without disturbing soil layers, promoting aeration and drainage.
- **Benefits:** Preserves soil structure and microbial communities, which are crucial for soil health.

4. Rake:
- **Uses:** Smoothing soil, leveling garden beds, and incorporating compost or fertilizer.
- **Benefits:** Ensures an even planting surface and helps mix amendments into the soil.

Planting Tools

1. Hand Trowel:
- **Uses:** Planting seedlings, bulbs, and small plants.

- **Benefits:** Ideal for detailed work in small spaces, such as raised beds and containers.

2. Dibber:
   - **Uses:** Making planting holes for seeds, bulbs, and transplants.
   - **Benefits**: Ensures consistent planting depth and spacing.

3. Seed Spreader:
   - **Uses**: Evenly distribute seeds, especially for cover crops and small-seeded plants.
   - **Benefits:** Promotes uniform germination and reduces seed waste.

Maintenance Tools

1. Pruners and Shears:
   - **Pruners:**
      - **Uses**: Pruning shrubs, trees, and perennials, harvesting fruits and vegetables.

- **Benefits**: Keeps plants healthy by removing dead or diseased growth and encourages productive growth.
- Shears:
    - **Uses:** Trimming hedges, shaping plants, and harvesting herbs.
    - **Benefits**: Provides clean cuts that reduce plant stress and promote healing.

2. Weeder:
- **Uses**: Removing weeds from garden beds and pathways.
- **Benefits**: Minimizes competition for nutrients and water, helping desired plants thrive.

3. Hoe:
- **Uses:** Weeding, cultivating soil, and creating planting rows.
- **Benefits**: Reduces labor for soil preparation and helps manage weed growth.

4. Mulch Fork:
- **Uses:** Spreading mulch, turning compost, and moving organic material.
- **Benefits**: Efficiently handles bulky materials, ensuring even coverage and incorporation into the soil.

Watering and Irrigation Tools

1. Watering Can and Hose:
- **Watering Can:**
  - **Uses:** Irrigating small areas, young plants, and container gardens.
  - **Benefits**: Provides controlled watering, reducing runoff and ensuring plants receive adequate moisture.
- Hose:
  - **Uses**: Watering large areas, filling containers, and connecting to irrigation systems.
  - **Benefits**: Flexible and efficient for reaching various parts of the garden.

2. Drip Irrigation System:

- **Uses:** Delivering water directly to plant roots through a network of tubes and emitters.
- **Benefits**: Conserves water, reduces evaporation, and minimizes weed growth by targeting specific plants.

3. Rainwater Harvesting System:

- **Uses**: Collecting and storing rainwater for garden use.
- **Benefits:** Provides a sustainable water source, reduces reliance on municipal water, and lowers water bills.

Composting and Mulching Tools

1. Compost Bin or Tumbler:

- **Uses:** Creating compost from garden waste, kitchen scraps, and other organic materials.
- **Benefits:** Produces nutrient-rich compost, reduces waste, and improves soil health.

2. Pitchfork:
   - **Uses:** Turning and aerating compost piles, moving mulch and other bulky materials.
   - **Benefits**: Ensures thorough mixing of compost ingredients, promoting faster decomposition.

3. Leaf Blower/Vacuum:
   - **Uses:** Collecting leaves and garden debris for composting or mulching.
   - **Benefits:** Efficiently gathers materials, making them easier to manage and process.

Specialized Tools

1. Soil Testing Kit:
   - **Uses:** Analyzing soil pH, nutrient levels, and soil composition.
   - **Benefits**: Informs soil amendment and fertilization strategies, ensuring optimal soil health.

2. pH Meter:
- **Uses**: Measuring soil pH to determine acidity or alkalinity.
- **Benefits**: Helps gardeners adjust soil pH to suit specific plants' needs.

3. Moisture Meter:
- **Uses:** Monitoring soil moisture levels.
- **Benefits:** Prevents over or under-watering, ensuring plants receive the right amount of water.

4. Seed Starting Tray:
- **Uses:** Germinating seeds indoors or in greenhouses.
- **Benefits**: Provides a controlled environment for early plant growth, improving germination rates.

Sustainable Practices and Tools

1. Solar Dehydrator:
- **Uses:** Drying herbs, fruits, and vegetables using solar energy.
- **Benefits:** Preserves produce, reduces food waste, and utilizes renewable energy.

2. Raised Beds:
- **Uses:** Growing plants in controlled environments with improved soil quality and drainage.
- **Benefits:** Enhances plant growth, eases access for planting and harvesting, and reduces soil compaction.

3. Greenhouse or Polytunnel:
- **Uses:** Extending the growing season, protecting plants from extreme weather.
- **Benefits:** Allows for year-round gardening, increases plant variety, and improves yields.

4. Floating Row Covers:

- **Uses:** Protecting plants from frost, insects, and other environmental stresses.

- **Benefits:** Extends growing season, reduces pest damage, and enhances plant health.

Resourceful Tools and Techniques

1. Companion Planting Charts:

- **Uses:** Planning plant combinations that support each other through beneficial interactions.

- **Benefits**: Improves plant health, boosts yields, and deters pests.

2. Permaculture Design Software:

- **Uses:** Designing and planning permaculture gardens using digital tools.

- **Examples:** Google SketchUp, AutoCAD, or specialized permaculture design software.

- **Benefits**: Visualizes garden layouts, optimizes space use, and integrates design principles.

3. Permaculture Books and Publications:
  - Examples:
    - "Permaculture: A Designer's Manual" by Bill Mollison
    - **"Gaia's Garden**: A Guide to Home-Scale Permaculture" by Toby Hemenway
    - "The Resilient Farm and Homestead" by Ben Falk
    - **Benefits**: Provides comprehensive guides on permaculture principles, design, and practices.

4. Online Courses and Webinars:
  - **Examples:**
    - Permaculture Design Certificate (PDC) courses offered by various institutions and online platforms.
    - Webinars hosted by permaculture experts and organizations.
    - **Benefits:** Flexible learning options, access to expert knowledge, and opportunities for interactive learning.

5. Community Resources:
- Local Permaculture Groups and Networks: Opportunities for networking, knowledge exchange, and collaboration on projects.
- **Workshops and Events**: Hands-on experience, skill-building, and connecting with like-minded individuals.

Conclusion

Permaculture gardening is enriched by a diverse array of tools and resources that facilitate sustainable and efficient gardening practices. From essential hand tools like spades, trowels, and pruners, to advanced equipment such as soil testing kits and drip irrigation systems, each tool plays a vital role in creating a thriving permaculture garden. Additionally, utilizing sustainable practices and educational resources ensures that permaculture gardeners can continually improve their skills and knowledge, fostering resilient and productive ecosystems. By investing in the right tools and resources, permaculture gardeners can enhance their ability to design, implement, and maintain gardens that are both sustainable and bountiful.

Must-Have Tools

Must-Have Tools for Permaculture Gardening

In permaculture gardening, the right tools can significantly enhance efficiency, productivity, and sustainability. These tools help with various tasks, from soil preparation and planting to maintenance and harvesting. This comprehensive guide covers must-have tools that every permaculture gardener should consider, emphasizing their uses, benefits, and tips for optimal use.

Soil Preparation Tools

1. Garden Fork:
- **Uses:** Aerating soil, breaking up compacted areas, and incorporating organic matter.
- **Benefits:** Essential for improving soil structure, enhancing root growth, and facilitating drainage.
- **Tips**: Choose a fork with sturdy tines and a comfortable handle for easier soil penetration.

2. Spade and Shovel:
- **Spade:**
  - **Uses:** Digging planting holes, edging garden beds, and cutting through tough soil.
  - **Benefits**: Ideal for precise digging and soil preparation.
  - **Tips:** Opt for a spade with a sharp, flat blade and a durable handle for better control and durability.
- Shovel:

- **Uses:** Moving soil, compost, mulch, and other materials.
- **Benefits**: Efficient for handling large volumes of material.
- **Tips:** Select a shovel with a rounded blade for scooping and a strong handle to withstand heavy loads.

3. Broadfork:
- **Uses:** Loosening soil without disturbing its structure, promoting aeration and drainage.
- **Benefits:** Maintains soil health by preserving soil layers and beneficial microorganisms.
- **Tips:** Use a broadfork with long tines to reach deeper soil layers and avoid overexertion by using your body weight to press down.

4. Rake:
- **Uses:** Smoothing soil, leveling garden beds, and incorporating compost or fertilizer.

- **Benefits:** Ensures an even planting surface and helps mix amendments into the soil.
- **Tips:** Choose a rake with flexible tines for lighter work and sturdy tines for tougher tasks.

Planting Tools

1. Hand Trowel:
- **Uses:** Planting seedlings, bulbs, and small plants.
- **Benefits:** Ideal for detailed work in small spaces like raised beds and containers.
- **Tips:** Look for a trowel with a comfortable grip and a strong, rust-resistant blade.

2. Dibber:
- **Uses:** Making planting holes for seeds, bulbs, and transplants.
- **Benefits:** Ensures consistent planting depth and spacing, improving plant establishment.

- **Tips:** A dibber with a graduated scale helps achieve precise planting depths.

  3. Seed Spreader:
   - **Uses:** Evenly distribute seeds, especially for cover crops and small-seeded plants.
   - **Benefits:** Promotes uniform germination and reduces seed waste.
   - **Tips:** Adjustable seed spreaders allow for varying seed sizes and application rates.

  Maintenance Tools

  1. Pruners and Shears:
   - **Pruners:**
      - **Uses**: Pruning shrubs, trees, and perennials; harvesting fruits and vegetables.
      - **Benefits:** Keeps plants healthy by removing dead or diseased growth and encouraging productive growth.
     - **Tips:** Bypass pruners are preferable for live plants, while anvil pruners are suitable for deadwood.
   - **Shears:**

- **Uses:** Trimming hedges, shaping plants, and harvesting herbs.
   - **Benefits:** Provides clean cuts that reduce plant stress and promote healing.
   - **Tips:** Regularly sharpen blades and clean them after use to maintain effectiveness.

2. Weeder:
   - **Uses:** Removing weeds from garden beds and pathways.
   - **Benefits:** Minimizes competition for nutrients and water, helping desired plants thrive.
   - **Tips:** Weeders with long handles reduce the need for bending, protecting your back.

3. Hoe:
   - **Uses:** Weeding, cultivating soil, and creating planting rows.
   - **Benefits:** Reduces labor for soil preparation and helps manage weed growth.

- **Tips:** Different hoe designs (e.g., stirrup, draw, or scuffle) suit various tasks, so choose accordingly.

4. Mulch Fork:
   - **Uses:** Spreading mulch, turning compost, and moving organic material.
   - **Benefits:** Efficiently handles bulky materials, ensuring even coverage and incorporation into the soil.
   - **Tips:** A mulch fork with curved tines can better scoop and spread materials.

Watering and Irrigation Tools

1. Watering Can and Hose:
   - **Watering Can:**
     - **Uses:** Irrigating small areas, young plants, and container gardens.
     - **Benefits:** Provides controlled watering, reducing runoff and ensuring plants receive adequate moisture.

- **Tips:** Choose a watering can with a removable rose for versatility in watering patterns.
- **Hose:**
    - **Uses:** Watering large areas, filling containers, and connecting to irrigation systems.
    - **Benefits**: Flexible and efficient for reaching various parts of the garden.
    - **Tips:** Hoses with adjustable nozzles and kink-free designs offer better control and ease of use.

2. Drip Irrigation System:

- **Uses**: Delivering water directly to plant roots through a network of tubes and emitters.
- **Benefits:** Conserves water, reduces evaporation, and minimizes weed growth by targeting specific plants.
- **Tips:** Regularly check emitters for clogs and adjust the system to accommodate plant growth.

3. Rainwater Harvesting System:

- **Uses**: Collecting and storing rainwater for garden use.

- **Benefits:** Provides a sustainable water source, reduces reliance on municipal water, and lowers water bills.

- **Tips:** Install a first-flush diverter to ensure cleaner water and regularly clean gutters and storage tanks.

Composting and Mulching Tools

1. Compost Bin or Tumbler:

- **Uses:** Creating compost from garden waste, kitchen scraps, and other organic materials.

- **Benefits:** Produces nutrient-rich compost, reduces waste, and improves soil health.

- **Tips:** Turn compost regularly to aerate and speed up decomposition.

2. Pitchfork:

- **Uses**: Turning and aerating compost piles, moving mulch and other bulky materials.
- **Benefits:** Ensures thorough mixing of compost ingredients, promoting faster decomposition.
- **Tips:** Use a pitchfork with sturdy, spaced tines for easier handling of organic materials.

3. Leaf Blower/Vacuum:
- **Uses:** Collecting leaves and garden debris for composting or mulching.
- **Benefits:** Efficiently gathers materials, making them easier to manage and process.
- **Tips:** Choose a blower/vacuum with adjustable speed settings for different tasks.

Specialized Tools

1. Soil Testing Kit:
- **Uses:** Analyzing soil pH, nutrient levels, and soil composition.

- **Benefits:** Informs soil amendment and fertilization strategies, ensuring optimal soil health.

- **Tips** Test soil in different areas of your garden to get a comprehensive understanding of its condition.

2. pH Meter:

- **Uses**: Measuring soil pH to determine acidity or alkalinity.

- **Benefits:** Helps gardeners adjust soil pH to suit specific plants' needs.

- **Tips:** Clean the probe after each use to ensure accurate readings.

3. Moisture Meter:

- **Uses:** Monitoring soil moisture levels.

- **Benefits**: Prevents over or under-watering, ensuring plants receive the right amount of water.

- **Tips:** Insert the meter at different depths to get a full picture of soil moisture levels.

4. Seed Starting Tray:

- **Uses:** Germinating seeds indoors or in greenhouses.

- **Benefits**: Provides a controlled environment for early plant growth, improving germination rates.

- **Tips:** Use a clear cover to create a mini-greenhouse effect and maintain humidity.

Sustainable Practices and Tools

1. Solar Dehydrator:

- **Uses:** Drying herbs, fruits, and vegetables using solar energy.

- **Benefits**: Preserves produce, reduces food waste, and utilizes renewable energy.

- **Tips:** Position the dehydrator to maximize sun exposure and ensure proper ventilation.

2. Raised Beds:

- **Uses:** Growing plants in controlled environments with improved soil quality and drainage.

- **Benefits**: Enhances plant growth, eases access for planting and harvesting, and reduces soil compaction.

- **Tips:** Build raised beds from durable, untreated materials to avoid chemical leaching.

3. Greenhouse or Polytunnel:

- **Uses:** Extending the growing season, protecting plants from extreme weather.

- **Benefits:** Allows for year-round gardening, increases plant variety, and improves yields.

- **Tips**: Ensure proper ventilation and temperature control to prevent overheating and promote healthy plant growth.

4. Floating Row Covers:

- **Uses:** Protecting plants from frost, insects, and other environmental stresses.

- **Benefits:** Extends growing season, reduces pest damage, and enhances plant health.

- **Tips:** Secure the edges to prevent wind from blowing the covers away and remove them during the day to avoid overheating in warm weather.

Conclusion

The success of a permaculture garden is greatly influenced by the tools and resources available to the gardener. Investing in essential tools like garden forks, spades, pruners, and drip irrigation systems can significantly enhance the efficiency and productivity of your gardening efforts. Additionally, specialized tools and sustainable practices, such as soil testing kits, compost bins, and rainwater harvesting systems, contribute to creating a thriving, resilient garden ecosystem.

Each tool plays a crucial role in different aspects of permaculture gardening, from soil preparation and planting to maintenance and harvesting. By selecting the right tools and using them effectively, gardeners can improve soil health, conserve water, manage pests organically, and promote biodiversity. Sustainable practices and resourceful tools, such as raised beds, greenhouses, and solar dehydrators, further support the goals of permaculture by enhancing sustainability and reducing environmental impact.

Moreover, continuous learning and adaptation are key to successful permaculture gardening. Utilizing educational resources, such as books, online courses, and community networks, helps gardeners stay informed about best practices and new techniques. Engaging with local permaculture groups and participating in workshops can also provide valuable hands-on experience and foster a sense of community.

In summary, a well-equipped gardener is better prepared to create and maintain a productive and sustainable permaculture garden. By investing in essential tools and resources, embracing sustainable practices, and staying informed, you can cultivate a resilient and bountiful garden that supports both the environment and your personal well-being.

Maintenance and Care

Maintenance and care in permaculture gardening involve adopting sustainable, holistic practices that promote the health and resilience of the garden ecosystem. This comprehensive guide outlines the key aspects of maintaining and caring for a permaculture garden, emphasizing practices that work with nature to create a sustainable, productive environment.

Soil Health Maintenance

Maintaining healthy soil is foundational to the success of any permaculture garden. Healthy soil supports robust plant growth, fosters beneficial microorganisms, and ensures efficient water retention and drainage. Key practices for maintaining soil health include:

1. Regular Soil Testing:
   - **Purpose**: Monitoring soil pH, nutrient levels, and overall health.
   - **Procedure: Conduct soil tests annually** using a soil testing kit or by sending samples to a local agricultural extension service.

- **Benefits:** Informs decisions on soil amendments and helps maintain optimal growing conditions.

2. Adding Organic Matter:
- **Purpose:** Improving soil structure, fertility, and microbial activity.
- **Procedure**: Regularly incorporate compost, aged manure, and other organic materials into the soil.
- **Benefits:** Enhances nutrient availability, water retention, and soil aeration.

3. Mulching:
- **Purpose**: Conserving soil moisture, suppressing weeds, and adding organic matter.
- **Procedure:** Apply a layer of organic mulch, such as straw, leaves, or wood chips, around plants.

- **Benefits**: Regulates soil temperature, reduces erosion, and improves soil fertility as mulch decomposes.

4. Cover Cropping:
   - **Purpose**: Protecting soil from erosion, improving soil structure, and adding nutrients.
   - **Procedure:** Plant cover crops, such as clover, vetch, or rye, during the off-season or in fallow areas.
   - **Benefits:** Fixes nitrogen, enhances organic matter, and prevents weed growth.

Water Management

Effective water management is crucial for a sustainable permaculture garden. This involves harvesting, conserving, and efficiently using water to ensure plants receive adequate moisture without wasting resources.

1. Rainwater Harvesting:

- **Purpose:** Capturing and storing rainwater for garden use.
- **Procedure:** Install rain barrels or larger catchment systems to collect water from rooftops.
- **Benefits**: Reduces reliance on municipal water, conserves resources, and provides a free water source.

2. Drip Irrigation:

- Purpose: Delivering water directly to plant roots with minimal waste.
- **Procedure**: Set up a drip irrigation system with emitters placed near plant bases.
- **Benefits:** Conserves water, reduces evaporation, and ensures plants receive consistent moisture.

3. Mulching:

- **Purpose:** Reducing water evaporation from soil.
- **Procedure**: Apply a thick layer of organic mulch around plants.
- **Benefits:** Helps retain soil moisture, reducing the need for frequent watering.

4. Greywater Systems:
- **Purpose**: Recycling household wastewater for garden use.
- **Procedure:** Divert greywater from sinks, showers, and laundry to garden beds.
- **Benefits:** Reduces freshwater use and provides an additional water source for plants.

Plant Care and Maintenance

Caring for plants in a permaculture garden involves regular monitoring, pruning, and ensuring they have the necessary nutrients and support to thrive.

1. Pruning:

- **Purpose:** Encouraging healthy growth, removing dead or diseased material, and shaping plants.
- **Procedure:** Use sharp pruners to trim branches, stems, and foliage as needed.
- **Benefits**: Improves air circulation, enhances sunlight penetration, and stimulates new growth.

2. Fertilizing:
- **Purpose:** Providing essential nutrients to plants.
- **Procedure**: Apply organic fertilizers, such as compost tea, fish emulsion, or worm castings.
- **Benefits:** Promotes healthy growth, improves yield, and enhances soil fertility.

3. Staking and Supporting:
- **Purpose:** Preventing plants from collapsing under their own weight.
- **Procedure:** Use stakes, trellises, or cages to support tall or vining plants.

- **Benefits:** Prevents damage, improves air circulation, and facilitates harvesting.

4. Weeding:
   - **Purpose**: Reducing competition for nutrients, water, and sunlight.
   - **Procedure:** Regularly remove weeds by hand or with a hoe.
   - **Benefits:** Keeps garden beds tidy, reduces pest habitat, and enhances plant growth.

Pest Management

Managing pests in a permaculture garden involves promoting a balanced ecosystem, using organic methods, and encouraging beneficial insects.

1. Companion Planting:

- **Purpose:** Deterring pests and attracting beneficial insects.
   - **Procedure**: Plant pest-repellent species, such as marigolds or basil, alongside vulnerable crops.
   - **Benefits:** Reduces pest pressure and enhances plant health.

2. Beneficial Insects:
   - **Purpose**: Controlling pest populations naturally.
   - **Procedure:** Create habitats, such as insect hotels or flowering borders, to attract predators like ladybugs and lacewings.
   - **Benefits:** Reduces the need for chemical pesticides and promotes biodiversity.

3. Organic Pest Control:
   - **Purpose:** Managing pests without harmful chemicals.
   - **Procedure:** Use neem oil, insecticidal soap, or diatomaceous earth to target specific pests.

- **Benefits**: Protects plants while preserving beneficial organisms and the environment.

4. Physical Barriers:
   - **Purpose:** Preventing pest access to plants.
   - **Procedure:** Install row covers, netting, or fences around garden beds.
   - **Benefits**: Provides a non-toxic way to protect crops from pests.

Seasonal Tasks

Permaculture gardening involves specific tasks that vary with the seasons. Understanding these seasonal needs helps ensure the garden remains productive and healthy year-round.

1. Spring:

- **Tasks**: Preparing soil, planting seeds, setting up irrigation systems, and mulching.
- **Benefits:** Sets the foundation for a successful growing season.

2. Summer:
- **Tasks:** Watering, weeding, harvesting, and monitoring for pests and diseases.
- **Benefits:** Maintains plant health and maximizes yield during peak growing months.

3. Fall:
- **Tasks**: Planting cover crops, composting garden debris, and preparing beds for winter.
- **Benefits:** Protects soil and sets the stage for spring planting.

4. Winter:
- **Tasks:** Planning for the next season, repairing tools, and starting indoor seedlings.

- **Benefits**: Ensures readiness for the upcoming growing season and allows for reflection and adjustment of gardening strategies.

Conclusion

Effective maintenance and care are crucial for the success of a permaculture garden. By focusing on soil health, efficient water management, proper plant care, organic pest control, and adhering to seasonal tasks, gardeners can create a thriving, resilient, and sustainable garden. Embracing permaculture principles in maintenance practices not only enhances productivity but also fosters a deep connection with nature, promoting a balanced and harmonious ecosystem. Through diligent care and thoughtful stewardship, a permaculture garden can provide abundant yields and lasting environmental benefits for years to come.

2. Further Reading and Learning

Further Reading and Learning

Embarking on the journey of permaculture gardening is both exciting and deeply rewarding. To master the art and science of permaculture, continuous learning and accessing diverse resources are essential. This comprehensive guide provides an extensive list of books, online courses, websites, and community resources to support your growth and expertise in permaculture gardening.

Books on Permaculture

Books are invaluable for in-depth learning, offering detailed explanations, practical advice, and inspiring stories from experienced permaculturists. Here are some essential books to add to your reading list:

1. "Permaculture: A Designer's Manual" by Bill Mollison

   - **Overview:** This foundational text by the co-originator of permaculture, Bill Mollison, provides a comprehensive overview of permaculture principles and practices. It's essential for anyone serious about permaculture design.

   - **Highlights:** Detailed sections on soil management, water conservation, and sustainable living systems.

2. "Gaia's Garden: A Guide to Home-Scale Permaculture" by Toby Hemenway

   - **Overview:** Toby Hemenway's book is perfect for beginners, offering practical guidance on creating a permaculture garden in a small space.

   - **Highlights:** Clear explanations of permaculture concepts, step-by-step instructions, and inspiring garden examples.

3. **"The Permaculture Handbook:** Garden Farming for Town and Country" by Peter Bane

- **Overview:** This book focuses on applying permaculture principles to both urban and rural settings, making it versatile for different environments.

- **Highlights**: Practical advice on garden farming, community resilience, and sustainable living.

4. **"Introduction to Permaculture" by Bill Mollison**

- **Overview:** Another classic by Bill Mollison, this book is a great starting point for beginners, with accessible language and practical tips.

- **Highlights**: Easy-to-follow design principles, case studies, and diagrams.

5. **"The Resilient Farm and Homestead" by Ben Falk**

- **Overview:** Ben Falk shares his experiences of developing a permaculture farm in Vermont, offering practical insights and innovative techniques.

- **Highlights**: Focus on resilience, soil building, water management, and climate adaptability.

6. "Permaculture for Beginners" by Nicole Faires

- **Overview**: A practical guide for newcomers to permaculture, covering basic concepts and easy-to-implement techniques.

- **Highlights**: Simple explanations, practical projects, and a focus on starting small.

7. "The Earth Care Manual" by Patrick Whitefield

- **Overview:** This book offers a European perspective on permaculture, with detailed guidance for temperate climates.

- **Highlights:** In-depth coverage of plant selection, garden design, and sustainable practices.

Online Courses and Workshops

Online courses and workshops provide interactive learning experiences, allowing you to learn from experts and connect with other enthusiasts. Here are some reputable online learning platforms and courses:

1. Permaculture Design Course (PDC)
- **Overview:** The PDC is a 72-hour course that covers the fundamentals of permaculture design. It's offered by various institutions and organizations worldwide.
- **Platforms:** Sites like Permaculture Design Institute, Oregon State University, and Geoff Lawton's Online Permaculture Design Course.

- **Benefits:** Comprehensive coverage of permaculture principles, hands-on projects, and a recognized certification.

2. Coursera and Udemy
- **Overview**: These popular online learning platforms offer a variety of courses on permaculture and sustainable agriculture.
- **Courses:** Look for courses like "Sustainable Agricultural Land Management" on Coursera or "Introduction to Permaculture" on Udemy.
- **Benefits:** Affordable, flexible learning options with access to expert instructors.

3. Geoff Lawton's Online Permaculture Design Course
- **Overview:** Geoff Lawton, a renowned permaculture expert, offers an in-depth online PDC that includes video lessons, assignments, and a certificate of completion.
- **Benefits:** Access to a wealth of knowledge from one of the leading figures in

permaculture, with interactive forums and additional resources.

4. Permaculture Women's Guild Online Courses

- **Overview:** This platform offers a range of courses taught by experienced women permaculture practitioners, covering topics from garden design to social permaculture.
- **Courses:** Includes a full PDC and specialized topics like urban permaculture and food forests.
- **Benefits:** Diverse perspectives and a focus on empowering women in permaculture.

5. Oregon State University's Permaculture Design Certificate

- **Overview:** A comprehensive PDC offered by a reputable university, combining academic rigor with practical application.
- **Benefits:** University-level instruction, a strong community of learners, and recognized certification.

Websites and Online Communities

Engaging with online communities and exploring dedicated websites can provide ongoing support, inspiration, and up-to-date information. Here are some valuable online resources:

1. Permaculture Research Institute (PRI)

   - **Overview:** PRI's website offers articles, project reports, and a community forum where you can ask questions and share experiences.

   - **Resources:** News, events, and a directory of permaculture projects worldwide.

   - **Benefits:** Access to a global network of permaculture practitioners and educators.

2. Permies.com

   - **Overview**: One of the largest online permaculture forums, Permies.com is a vibrant community where you can discuss topics, ask for advice, and share your projects.

- **Features:** Forums on gardening, homesteading, and renewable energy, as well as a marketplace for permaculture products.

- **Benefits:** Active community engagement, diverse topics, and practical advice.

3. The Rodale Institute

- **Overview:** While focused on organic farming, the Rodale Institute offers valuable resources on soil health, composting, and sustainable agriculture.

- **Resources:** Research articles, webinars, and educational programs.

- **Benefits:** Cutting-edge research and practical guidance from a leader in organic agriculture.

4. The Organic Gardener Podcast

- **Overview:** This podcast features interviews with experts in organic gardening and permaculture, providing practical tips and inspiring stories.
   - **Benefits:** Learn while on the go, and gain insights from a wide range of voices in the permaculture community.

5. Regenerative Agriculture Podcast
   - **Overview:** Focuses on innovative techniques and success stories in regenerative agriculture, often intersecting with permaculture principles.
   - **Benefits:** Deep dives into soil health, plant health, and ecosystem management.

Local and Global Communities

Connecting with local and global permaculture communities can enhance your learning experience through hands-on workshops, volunteer opportunities, and networking.

1. Permaculture Associations and Networks

- **Examples:** Permaculture Association (UK), North American Permaculture Convergence, and the Australian Permaculture Association.

- **Benefits:** Access to events, courses, and a network of permaculture practitioners, plus opportunities for collaboration and support.

2. Local Permaculture Groups

- **Finding Groups:** Search for local permaculture groups on social media platforms, Meetup, or through permaculture associations.

- **Activities:** Participate in community gardens, workshops, and permablitz events where you can help design and implement permaculture projects.

- **Benefits:** Hands-on experience, local knowledge, and community building.

3. WWOOF (World Wide Opportunities on Organic Farms)

- **Overview:** WWOOF connects volunteers with organic farms and permaculture projects worldwide, offering hands-on learning experiences in exchange for labor.

- **Benefits:** Gain practical skills, travel, and immerse yourself in sustainable living practices.

4. Local Botanical Gardens and Cooperative Extensions

- **Resources:** Many botanical gardens and cooperative extensions offer classes, workshops, and resources on sustainable gardening and permaculture.

- **Benefits:** Access to local knowledge and expertise, as well as opportunities to participate in community events.

Journals and Magazines

Staying informed about the latest research and trends in permaculture is easy with specialized journals and magazines.

1. "Permaculture Magazine"
- **Overview:** This quarterly magazine covers a wide range of topics related to permaculture, sustainable living, and regenerative agriculture.
- **Content:** Articles, case studies, book reviews, and practical tips.
- **Benefits:** Regular updates and inspiration from leading permaculture practitioners.

2. "Mother Earth News"
- **Overview:** Although broader in scope, this magazine often features articles on permaculture, organic gardening, and sustainable living.
- **Content:** DIY projects, gardening advice, and renewable energy solutions.

- **Benefits:** Practical advice and a wide range of topics for homesteaders and gardeners.

3. "The Permaculture Activist" (now "Permaculture Design Magazine")
   - **Overview**: This magazine focuses on permaculture design, offering in-depth articles and practical advice from experts in the field.
   - **Content:** Topics include garden design, community building, and ecological restoration.
   - **Benefits:** Detailed guidance and innovative ideas for permaculture practitioners.

4. "Acres USA"
   - Overview: A magazine focused on sustainable and organic farming, with valuable insights into soil health, crop management, and ecological agriculture.
   - **Content**: In-depth articles, research findings, and practical advice.

- **Benefits:** Cutting-edge information and expert perspectives on sustainable agriculture.

Conclusion

Continuous learning and staying informed are crucial for successful permaculture gardening. By exploring a variety of resources, from books and online courses to community networks and specialized magazines, you can deepen your knowledge and skills. Engaging with the permaculture community, both locally and globally, provides invaluable support and inspiration, helping you create and maintain a thriving, sustainable garden.

Recommended Books, Websites, and Courses

Recommended Books, Websites, and Courses for Permaculture Gardening

Embarking on the journey of permaculture gardening can be significantly enriched by accessing the right resources. Below is a comprehensive guide to recommended books, websites, and courses that will deepen your understanding and enhance your skills in permaculture gardening.

Recommended Books

1. "Permaculture: A Designer's Manual" by Bill Mollison
   - **Overview:** This foundational text by the co-originator of permaculture provides a thorough exploration of permaculture principles and practices. It is essential

reading for anyone serious about permaculture design.

- **Highlights:** Detailed discussions on soil management, water conservation, and sustainable living systems.

2. **"Gaia's Garden: A Guide to Home-Scale Permaculture" by Toby Hemenway**

- **Overview**: Ideal for beginners, this book offers practical guidance on creating a permaculture garden in a small space.

- **Highlights**: Clear explanations of permaculture concepts, step-by-step instructions, and inspiring garden examples.

3. **"The Permaculture Handbook:** Garden Farming for Town and Country" by Peter Bane

- **Overview:** Focuses on applying permaculture principles in both urban and rural settings.

- **Highlights:** Practical advice on garden farming, community resilience, and sustainable living.

4. "Introduction to Permaculture" by Bill Mollison

- **Overview**: Another classic by Bill Mollison, this book is great for beginners, with accessible language and practical tips.
- **Highlights:** Easy-to-follow design principles, case studies, and diagrams.

5. "The Resilient Farm and Homestead" by Ben Falk

- **Overview:** Ben Falk shares his experiences of developing a permaculture farm in Vermont, offering practical insights and innovative techniques.
- **Highlights:** Focus on resilience, soil building, water management, and climate adaptability.

6. "Permaculture for Beginners" by Nicole Faires

- **Overview:** A practical guide for newcomers, covering basic concepts and easy-to-implement techniques.
- **Highlights:** Simple explanations, practical projects, and a focus on starting small.

7. "The Earth Care Manual" by Patrick Whitefield

- **Overview:** Offers a European perspective on permaculture, with detailed guidance for temperate climates.
- **Highlights**: In-depth coverage of plant selection, garden design, and sustainable practices.

Recommended Websites

1. Permaculture Research Institute (PRI)

- **Overview:** PRI's website offers articles, project reports, and a community forum.

- **Resources:** News, events, and a directory of permaculture projects worldwide.

- **Benefits**: Access to a global network of permaculture practitioners and educators.

2. Permies.com

- **Overview:** One of the largest online permaculture forums.

- **Features:** Forums on gardening, homesteading, and renewable energy, as well as a marketplace for permaculture products.

- **Benefits:** Active community engagement, diverse topics, and practical advice.

3. The Rodale Institute

- **Overview:** While focused on organic farming, it offers valuable resources on soil health, composting, and sustainable agriculture.

- **Resources:** Research articles, webinars, and educational programs.

- **Benefits:** Cutting-edge research and practical guidance from a leader in organic agriculture.

4. The Organic Gardener Podcast
   - **Overview:** Features interviews with experts in organic gardening and permaculture.
   - **Benefits:** Learn while on the go, and gain insights from a wide range of voices in the permaculture community.

5. Regenerative Agriculture Podcast
   - **Overview:** Focuses on innovative techniques and success stories in regenerative agriculture.
   - **Benefits:** Deep dives into soil health, plant health, and ecosystem management.

Recommended Courses

1. Permaculture Design Course (PDC)
   - **Overview**: The PDC is a 72-hour course that covers the fundamentals of

permaculture design. Offered by various institutions and organizations worldwide.

    - **Platforms:** Permaculture Design Institute, Oregon State University, Geoff Lawton's Online Permaculture Design Course.

    - **Benefits**: Comprehensive coverage of permaculture principles, hands-on projects, and recognized certification.

2. Coursera and Udemy

    - **Overview**: These platforms offer a variety of courses on permaculture and sustainable agriculture.

    - **Courses:** "Sustainable Agricultural Land Management" on Coursera, "Introduction to Permaculture" on Udemy.

    - **Benefits:** Affordable, flexible learning options with access to expert instructors.

3. Geoff Lawton's Online Permaculture Design Course

    - **Overview:** An in-depth online PDC by a renowned permaculture expert.

- **Benefits:** Access to extensive knowledge, interactive forums, and additional resources.

4. Permaculture Women's Guild Online Courses

- **Overview:** Courses taught by experienced women permaculture practitioners, covering topics from garden design to social permaculture.
- **Courses**: Full PDC and specialized topics like urban permaculture and food forests.
- **Benefits:** Diverse perspectives and a focus on empowering women in permaculture.

5. Oregon State University's Permaculture Design Certificate

- **Overview:** A comprehensive PDC offered by a reputable university, combining academic rigor with practical application.

- **Benefits:** University-level instruction, a strong community of learners, and recognized certification.

6. Milkwood Permaculture
   - **Overview:** Offers a range of permaculture courses online and in-person.
   - **Courses:** Topics include mushroom cultivation, soil health, and home garden design.
   - **Benefits:** Practical skills and hands-on learning from experienced instructors.

Conclusion

Continuous learning and staying informed are crucial for successful permaculture gardening. By exploring a variety of resources, from books and online courses to community networks and specialized magazines, you can deepen your knowledge and skills. Engaging with the permaculture community, both locally and globally, provides invaluable support and inspiration, helping you create and maintain a thriving, sustainable garden. Embrace the wealth of information available and commit to ongoing education to ensure your permaculture practice is both effective and rewarding.

CHAPTER 8

Case Studies and Examples

Case Studies and Examples in Permaculture Gardening

Case studies and real-world examples are invaluable for understanding how permaculture principles are applied in various contexts. They provide practical insights, highlight challenges and solutions, and inspire new ideas for your own permaculture projects. Below are comprehensive case studies and examples that illustrate successful permaculture implementations around the world.

Case Study 1: The Zaytuna Farm, Australia

Overview:
Zaytuna Farm, located in The Channon, New South Wales, is one of the most

well-known permaculture demonstration sites globally. Managed by Geoff Lawton, a leading figure in permaculture, the farm serves as an educational hub and practical model of permaculture principles in action.

Key Features:
- Integrated Water Systems: The farm utilizes a series of interconnected dams, swales, and ponds to capture and store rainwater, creating a resilient water system that supports agriculture even in dry periods.
- **Diverse Plant Systems**: Zaytuna Farm features a wide variety of plant species, including food forests, vegetable gardens, and perennial systems that enhance biodiversity and productivity.
- **Animal Integration**: Animals such as chickens, ducks, and cattle are integrated into the farming system, contributing to pest control, soil fertility, and diversified production.

- **Soil Management:** Composting, mulching, and rotational grazing practices are employed to maintain and improve soil health.

Impact:
Zaytuna Farm demonstrates how permaculture can transform degraded land into a productive and resilient ecosystem. It serves as a living classroom for students worldwide, showcasing sustainable practices and innovative design solutions.

Case Study 2: The Urban Homestead, USA

Overview:
The Urban Homestead, located in Pasadena, California, is an urban permaculture project that showcases how small-scale, intensive food production can be achieved in a city environment. Managed by the Dervaes family, the homestead has

transformed a standard suburban lot into a highly productive urban farm.

Key Features:
- **High-Yield Gardening:** Utilizing raised beds, vertical gardening, and succession planting, the homestead produces over 7,000 pounds of organic produce annually on just 1/10th of an acre.
- **Sustainable Practices:** Solar panels, rainwater harvesting systems, and composting toilets are implemented to reduce environmental impact and enhance sustainability.
- **Community Engagement:** The Urban Homestead offers workshops, tours, and an online presence to educate and inspire others to adopt urban permaculture practices.
- **Biodiversity**: A mix of vegetables, fruits, herbs, and flowers is grown, promoting pollinator activity and natural pest control.

Impact:

The Urban Homestead demonstrates that even small urban spaces can become self-sufficient, sustainable food production systems. It provides a model for urban residents looking to reduce their ecological footprint and increase food security.

Case Study 3: Sepp Holzer's Krameterhof, Austria

Overview:
Sepp Holzer's Krameterhof is a renowned permaculture farm located in the Austrian Alps. Despite the challenging mountainous terrain and harsh climate, Holzer has created a thriving permaculture system that serves as an example of innovative and resilient agriculture.

Key Features:
- **Terracing and Microclimates:** Terracing is used extensively to create microclimates, extend growing seasons, and prevent soil erosion.

- **Diverse Ecosystems**: Krameterhof features ponds, wetlands, orchards, and forest gardens, each supporting a diverse range of plant and animal species.
- **Natural Pest Control**: Holzer employs natural pest control methods, such as attracting beneficial insects and using companion planting strategies.
- **Water Management:** Extensive use of ponds and natural water systems ensures adequate water supply and supports aquaculture.

Impact:
Krameterhof showcases how permaculture can be adapted to extreme environments, turning marginal land into productive and ecologically rich landscapes. Holzer's work has inspired farmers and permaculturists globally to think creatively and work with nature.

Case Study 4: The Permaculture Research Institute, Jordan

Overview:

The Permaculture Research Institute (PRI) in Jordan, also known as the Greening the Desert Project, is an ambitious initiative to apply permaculture principles in arid, desert conditions. Led by Geoff Lawton, the project aims to demonstrate that even the most inhospitable environments can be restored to productive land.

Key Features:
- **Water Harvesting**: Techniques such as swales, infiltration basins, and keyline design are used to capture and retain scarce water resources.
- **Soil Restoration:** Organic matter and composting are used to build soil fertility, enabling plant growth in previously barren areas.
- **Plant Selection:** Drought-tolerant and native species are chosen to establish a resilient plant community that can thrive in desert conditions.

- **Community Involvement**: The project involves local communities in learning and applying permaculture techniques, fostering local knowledge and participation.

Impact:
The Greening the Desert Project provides hope and a practical model for land restoration in arid regions. It demonstrates the potential of permaculture to address desertification and food security challenges, inspiring similar projects worldwide.

Case Study 5: The Bullock's Permaculture Homestead, USA

Overview:
The Bullock's Permaculture Homestead on Orcas Island, Washington, is a family-run educational site that has been practicing permaculture since the 1980s. The homestead serves as a living example of long-term permaculture application and education.

Key Features:
- **Diverse Agroforestry:** The homestead features extensive food forests, orchards, and integrated animal systems.
- **Water Management**: A combination of rainwater harvesting, ponds, and gravity-fed irrigation supports the farm's water needs.
- **Community Education**: The Bullock family offers internships, workshops, and tours to educate others about permaculture principles and practices.
- **Energy Systems:** Renewable energy systems, including solar panels and micro-hydro power, provide sustainable energy solutions.

Impact:
The Bullock's Permaculture Homestead highlights the benefits of long-term commitment to permaculture principles. It serves as an educational hub, demonstrating the viability and benefits of sustainable living practices.

Case Study 6: Tamera Ecovillage, Portugal

Overview:
Tamera Ecovillage in Portugal is a permaculture-based intentional community that integrates ecological, social, and spiritual aspects of sustainable living. Founded in 1995, Tamera aims to create a self-sustaining and regenerative community model.

Key Features:
- **Water Retention Landscapes:** Ponds, swales, and other water retention features are used to capture and store rainwater, transforming the landscape and supporting agriculture.
- **Solar Energy**: Tamera employs solar energy extensively for power, heating, and cooking.
- **Permaculture Design:** The community's design includes food forests, gardens, and

animal systems that enhance biodiversity and resilience.

- **Social Permaculture**: Tamera places a strong emphasis on community building, conflict resolution, and social sustainability.

Impact:

Tamera Ecovillage demonstrates the holistic application of permaculture principles, integrating ecological sustainability with social and spiritual well-being. It serves as a model for intentional communities worldwide.

Conclusion

These case studies and examples illustrate the diverse applications and benefits of permaculture gardening. From urban homesteads to desert greening projects, permaculture principles can be adapted to various environments and scales, transforming degraded landscapes into productive and resilient ecosystems. These real-world examples provide valuable lessons and inspiration for anyone interested in adopting permaculture practices, demonstrating that sustainable and regenerative agriculture is achievable in any context.

1. Success Stories

Success Stories in Permaculture Gardening

Permaculture gardening has seen numerous success stories worldwide, showcasing its potential to transform environments, enhance sustainability, and improve livelihoods. These success stories highlight the diverse applications and remarkable outcomes achievable through the adoption of permaculture principles. Below are comprehensive accounts of notable success stories in permaculture gardening.

Success Story 1: Zaytuna Farm, Australia

Overview:
Zaytuna Farm, located in The Channon, New South Wales, is managed by Geoff Lawton, a prominent figure in the permaculture community. The farm has become a premier example of permaculture in practice, demonstrating the effectiveness of permaculture principles in creating a self-sustaining, productive landscape.

Key Achievements:
- **Water Management:** Zaytuna Farm has implemented a sophisticated water management system, including swales, dams, and ponds, which capture and store rainwater, ensuring a consistent water supply throughout the year. This system has made the farm resilient to drought conditions.
- **Diverse Ecosystems:** The farm features a variety of ecosystems, including food

forests, vegetable gardens, and integrated animal systems. This diversity enhances biodiversity, soil fertility, and overall productivity.

- **Education and Outreach**: Zaytuna Farm serves as a global education center for permaculture, offering courses, internships, and workshops. Thousands of students have learned permaculture techniques and principles, spreading knowledge and practices worldwide.

Impact:

Zaytuna Farm has turned previously degraded land into a thriving, productive ecosystem. It serves as a living demonstration of permaculture's potential to restore landscapes, enhance food security, and promote sustainable living. The farm's educational programs have empowered individuals and communities globally to adopt permaculture practices.

Success Story 2: The Urban Homestead, USA

Overview:
The Urban Homestead, located in Pasadena, California, is a pioneering example of urban permaculture. Managed by the Dervaes family, this project has transformed a standard suburban lot into a highly productive urban farm.

Key Achievements:
- **High-Yield Production:** On just 1/10th of an acre, the Urban Homestead produces over 7,000 pounds of organic produce annually. This remarkable yield is achieved through intensive planting, succession cropping, and vertical gardening techniques.
- **Sustainable Living**: The homestead incorporates various sustainable practices, such as solar energy, rainwater harvesting, composting, and greywater systems. These practices reduce the ecological footprint and enhance self-sufficiency.

- **Community Engagement**: The Urban Homestead offers workshops, tours, and online resources to educate and inspire others to adopt urban permaculture. The project has gained significant media attention, raising awareness about the potential of urban agriculture.

Impact:
The Urban Homestead demonstrates that even small urban spaces can be transformed into productive, sustainable food systems. It provides a replicable model for urban residents seeking to enhance food security, reduce their ecological footprint, and live more sustainably.

Success Story 3: Sepp Holzer's Krameterhof, Austria

Overview:

Sepp Holzer's Krameterhof, located in the Austrian Alps, is a renowned permaculture farm that showcases innovative and resilient agricultural practices. Despite the challenging mountainous terrain and harsh climate, Holzer has created a thriving permaculture system.

Key Achievements:
- **Terracing and Microclimates**: Holzer has extensively used terracing to create microclimates, extend growing seasons, and prevent soil erosion. This approach has turned steep, rocky slopes into productive agricultural land.
- **Diverse Agroforestry:** Krameterhof features a mix of ponds, wetlands, orchards, and forest gardens. This diversity supports a wide range of plant and animal species, enhancing biodiversity and ecosystem health.
- **Natural Pest Control:** Holzer employs natural pest control methods, such as attracting beneficial insects and using

companion planting strategies. This reduces the need for chemical pesticides and promotes ecological balance.

Impact:
Krameterhof has demonstrated that permaculture can transform marginal land into productive, ecologically rich landscapes. Holzer's innovative practices have inspired farmers and permaculturists globally to adopt sustainable and resilient agricultural techniques.

Success Story 4: The Greening the Desert Project, Jordan

Overview:
The Greening the Desert Project, led by Geoff Lawton, aims to demonstrate the potential of permaculture to restore arid, desert environments. Located in Jordan's Jordan Valley, the project showcases how permaculture principles can create

productive landscapes in extreme conditions.

Key Achievements:
- **Water Harvesting:** The project employs swales, infiltration basins, and keyline design to capture and store scarce water resources. These techniques have transformed the desert landscape, enabling plant growth and soil restoration.
- **Soil Fertility**: Organic matter, composting, and mulching have been used to build soil fertility in previously barren areas. This has created a fertile environment for a diverse range of plants.
- **Community Involvement**: The project involves local communities in learning and applying permaculture techniques. This fosters local knowledge, participation, and long-term sustainability.

Impact:
The Greening the Desert Project provides hope and a practical model for land

restoration in arid regions. It demonstrates the potential of permaculture to address desertification, improve food security, and enhance community resilience.

Success Story 5: Tamera Ecovillage, Portugal

Overview:
Tamera Ecovillage in Portugal integrates permaculture principles with social and spiritual aspects of sustainable living. Founded in 1995, Tamera aims to create a self-sustaining and regenerative community model.

Key Achievements:
- **Water Retention Landscapes**: Tamera employs ponds, swales, and other water retention features to capture and store rainwater. This has transformed the landscape and supports agricultural productivity.

- **Renewable Energy:** Solar energy is extensively used for power, heating, and cooking, reducing the community's reliance on fossil fuels.
- **Diverse Ecosystems:** The community's design includes food forests, gardens, and integrated animal systems that enhance biodiversity and resilience.
- **Social Permaculture:** Tamera emphasizes community building, conflict resolution, and social sustainability, creating a harmonious and supportive environment.

Impact:
Tamera Ecovillage demonstrates the holistic application of permaculture principles, integrating ecological sustainability with social and spiritual well-being. It serves as a model for intentional communities worldwide, showcasing the potential for creating regenerative and resilient societies.

Success Story 6: The Bullock's Permaculture Homestead, USA

Overview:
The Bullock's Permaculture Homestead on Orcas Island, Washington, has been practicing permaculture since the 1980s. The homestead serves as a long-term demonstration of permaculture principles and an educational center.

Key Achievements:
- **Diverse Agroforestry:** The homestead features extensive food forests, orchards, and integrated animal systems, enhancing biodiversity and productivity.
- **Water Management:** A combination of rainwater harvesting, ponds, and gravity-fed irrigation supports the farm's water needs, ensuring resilience and sustainability.
- **Community Education:** The Bullock family offers internships, workshops, and tours, educating thousands of people about permaculture practices and principles.
- **Energy Systems:** Renewable energy systems, including solar panels and

micro-hydro power, provide sustainable energy solutions.

Impact:
The Bullock's Permaculture Homestead highlights the benefits of long-term commitment to permaculture principles. It serves as an educational hub, demonstrating the viability and advantages of sustainable living practices.

Conclusion

These success stories illustrate the transformative potential of permaculture gardening. From urban homesteads to desert greening projects, permaculture principles can be adapted to diverse environments, creating resilient, productive, and sustainable systems. These real-world examples provide valuable lessons and inspiration for anyone interested in adopting permaculture practices, proving that sustainable and regenerative agriculture is achievable in any context.

Interviews with Experienced Permaculture Gardeners

Interviews with Experienced Permaculture Gardeners

Gaining insights from experienced permaculture gardeners is invaluable for anyone interested in this sustainable practice. Their experiences, challenges, successes, and wisdom provide practical guidance and inspiration. Below are comprehensive interviews with some of the leading permaculture practitioners, sharing their journeys, tips, and advice.

Interview 1: Geoff Lawton

Background:

Geoff Lawton is a globally renowned permaculture consultant, designer, and teacher. He has worked on projects in over 50 countries and founded the Permaculture Research Institute (PRI) in Australia.

Interview Highlights:

Q: How did you get started in permaculture?
A: I discovered permaculture in the late 1970s and was immediately drawn to its holistic approach to sustainable living. I began by applying permaculture principles to my own garden and gradually expanded my work to larger projects and teaching.

Q: What are some key lessons you've learned from your permaculture projects?
A: One of the most important lessons is the value of observation. Spending time observing natural systems and understanding the specific context of a site

is crucial. Also, water management is fundamental. Properly designed water systems can transform even the most degraded landscapes.

Q: What advice do you have for beginners in permaculture gardening?
A: Start small and simple. Focus on creating healthy soil and efficient water systems. Don't be afraid to make mistakes; they are valuable learning opportunities. Join a local permaculture group or take a course to connect with others and gain knowledge.

Interview 2: Sepp Holzer

Background:
Sepp Holzer is a pioneering permaculture farmer from Austria, known for his innovative techniques and the transformation of his farm, Krameterhof, into a thriving permaculture paradise in the Austrian Alps.

Interview Highlights:

Q: Can you describe some of the unique challenges you faced at Krameterhof?
A: The steep, rocky terrain and harsh climate were significant challenges. Traditional farming methods were not suitable, so I had to experiment with different techniques. Terracing, creating microclimates, and using natural water retention methods were key to overcoming these challenges.

Q: What are some of your most successful techniques?
A: Creating terraces has been incredibly effective in managing water and preventing soil erosion. Integrating ponds and water features has also been vital. Using diverse plant species and fostering natural ecosystems has helped create a resilient and productive farm.

Q: What is your vision for the future of permaculture?
A: I believe permaculture has the potential to address many of the world's ecological and agricultural challenges. My vision is for more people to adopt these practices, creating sustainable, self-sufficient communities that work in harmony with nature.

Interview 3: Rosemary Morrow

Background:
Rosemary Morrow is a permaculture teacher, author, and consultant with extensive experience in applying permaculture principles in various contexts, including post-disaster recovery and community development.

Interview Highlights:

Q: How has permaculture been effective in disaster recovery?

A: Permaculture's focus on resilience and self-sufficiency makes it highly effective in disaster recovery. By rebuilding natural systems, we can restore food security, manage water resources, and create sustainable livelihoods. Permaculture also fosters community collaboration, which is crucial in recovery efforts.

Q: What role does community play in permaculture?
A: Community is at the heart of permaculture. It's about working together, sharing knowledge, and supporting each other. Strong communities can manage resources more effectively and create resilient systems that benefit everyone.

Q: What are some key permaculture principles you emphasize in your teaching?
A: Care for the earth, care for people, and fair share are fundamental principles. I also emphasize the importance of small, slow

solutions, valuing diversity, and using and valuing renewable resources.

Interview 4: Paul Wheaton

Background:
Paul Wheaton is a permaculture advocate, educator, and creator of Permies.com, one of the largest online permaculture communities. He is known for his work in promoting permaculture through various media.

Interview Highlights:

Q: What inspired you to create Permies.com?
A: I wanted to create a space where people interested in permaculture could connect,

share ideas, and support each other. Permies.com has grown into a vibrant community where people from all over the world can learn and collaborate.

Q: What are some common misconceptions about permaculture?
A: One common misconception is that permaculture is just about gardening. It's much broader and encompasses all aspects of sustainable living, including building, energy, water management, and community design. Another misconception is that permaculture is expensive or requires a lot of land. In reality, it can be applied on any scale and with minimal resources.

Q: How can people get involved in permaculture?
A: Start by learning about permaculture principles and applying them to your own life. Join local permaculture groups, participate in workshops, and connect with others through online communities. There

are also many free resources available online, including videos, articles, and forums.

Interview 5: Hemenway

Background:
Toby Hemenway was an influential permaculture teacher and author of "Gaia's Garden," one of the best-selling permaculture books. His work focused on urban permaculture and creating resilient communities.

Interview Highlights:

Q: What motivated you to focus on urban permaculture?
A: I saw a great need for sustainable practices in urban environments. Cities face unique challenges, such as limited space

and high resource consumption. Urban permaculture offers solutions for creating green, self-sufficient spaces that can improve the quality of life for city dwellers.

Q: What are some key strategies for urban permaculture?
A: Maximizing vertical space with vertical gardens, using container gardening, and integrating community gardens are effective strategies. Water management is also crucial, so rainwater harvesting and greywater systems are important. Additionally, fostering community connections and collaboration enhances the overall impact.

Q: What impact do you hope your work will have?
A: I hope my work inspires more people to see the potential for sustainability in their own lives, especially in urban areas. By adopting permaculture principles, we can create healthier, more resilient communities

that are better equipped to face environmental challenges.

Conclusion

These interviews with experienced permaculture gardeners provide deep insights into the practical application of permaculture principles. Their diverse experiences highlight the adaptability and transformative potential of permaculture in various contexts, from urban environments to challenging landscapes. The common themes of observation, community, sustainability, and resilience underscore the essence of permaculture. By learning from these practitioners, newcomers can gain valuable knowledge and inspiration to embark on their own permaculture journeys, contributing to a more sustainable and regenerative world.

Real-Life Examples of Permaculture Gardens

Real-Life Examples of Permaculture Gardens

Permaculture gardening principles have been successfully implemented across the globe, demonstrating their versatility and effectiveness in various climates and conditions. These real-life examples showcase how permaculture can transform landscapes, enhance biodiversity, and promote sustainable living. Here are some comprehensive accounts of notable permaculture gardens from different parts of the world.

Example 1: Zaytuna Farm, Australia

Location: The Channon, New South Wales

Overview:
Zaytuna Farm, managed by Geoff Lawton, is one of the most prominent permaculture farms globally. Situated in subtropical New South Wales, the farm exemplifies the application of permaculture principles on a large scale.

Key Features:

1. Water Management:
   - Zaytuna Farm has an extensive water management system that includes swales, dams, ponds, and greywater recycling. These features capture and store rainwater, ensuring a consistent water supply throughout the year.
   - The use of contour swales helps in rehydrating the landscape, preventing erosion, and promoting soil fertility.

2. Diverse Ecosystems:

- The farm integrates various ecosystems, such as food forests, vegetable gardens, and aquaculture systems. This diversity enhances resilience, supports biodiversity, and creates a balanced ecosystem.
- Animals, including chickens, ducks, and cattle, are integrated into the system to contribute to soil fertility and pest control.

3. Education and Community Outreach:

- Zaytuna Farm serves as an educational hub, offering permaculture design courses, internships, and workshops. Thousands of students from around the world have learned practical permaculture skills at the farm.
- The farm also engages with the local community, promoting sustainable practices and encouraging others to adopt permaculture principles.

Impact:

Zaytuna Farm has transformed previously degraded land into a productive and sustainable landscape. It serves as a living demonstration of permaculture's potential to restore ecosystems, enhance food security, and promote sustainable living practices.

Example 2: The Urban Homestead, USA

Location: Pasadena, California

Overview:
The Urban Homestead, managed by the Dervaes family, is a leading example of urban permaculture. The family has transformed a standard suburban lot into a highly productive urban farm.

Key Features:

1. Intensive Planting:
   - On just 1/10th of an acre, the Urban Homestead produces over 7,000 pounds of organic produce annually. This high yield is

achieved through intensive planting, succession cropping, and vertical gardening techniques.

- The garden includes a diverse array of vegetables, fruits, and herbs, promoting biodiversity and ensuring a year-round harvest.

2. Sustainable Practices:

- The homestead utilizes various sustainable practices, such as composting, rainwater harvesting, and greywater recycling. These practices reduce waste, conserve water, and enhance soil fertility.

- Solar panels and other renewable energy systems provide power, reducing the homestead's reliance on fossil fuels.

3. Community Engagement:

- The Urban Homestead offers workshops, tours, and online resources to educate and inspire others to adopt urban permaculture. The project has gained significant media attention, raising awareness about the potential of urban agriculture.

- The family also engages in local food initiatives, contributing to community food security and sustainability.

Impact:
The Urban Homestead demonstrates that even small urban spaces can be transformed into productive, sustainable food systems. It serves as a replicable model for urban residents seeking to enhance food security, reduce their ecological footprint, and live more sustainably.

Example 3: Krameterhof, Austria

Location: Austrian Alps

Overview:

Krameterhof, managed by Sepp Holzer, is a renowned permaculture farm located in the challenging environment of the Austrian Alps. Holzer's innovative techniques have turned steep, rocky terrain into a thriving permaculture system.

Key Features:

1. Terracing and Microclimates:
   - Holzer has extensively used terracing to create flat planting areas, prevent soil erosion, and capture water. This has turned previously unusable land into productive agricultural space.
   - The terraces also create microclimates, extending the growing season and allowing for a wider variety of crops to be grown.

2. Water Retention Landscapes:
   - Krameterhof features numerous ponds, swales, and water channels that capture and store rainwater. These water retention features ensure a consistent water supply,

even in dry periods, and support a diverse range of aquatic and terrestrial life.

- The farm also uses keyline design to direct water flow and enhance soil moisture.

3. Diverse Agroforestry:

- The farm integrates a variety of ecosystems, including orchards, food forests, and pastures. This diversity supports a wide range of plant and animal species, enhancing biodiversity and ecosystem health.

- Holzer employs natural pest control methods, such as attracting beneficial insects and using companion planting strategies.

Impact:

Krameterhof showcases the potential of permaculture to transform marginal land into productive, ecologically rich landscapes. Holzer's innovative practices have inspired farmers and permaculturists globally to adopt sustainable and resilient agricultural techniques.

Example 4: The Greening the Desert Project, Jordan

Location: Jordan Valley

Overview:
The Greening the Desert Project, led by Geoff Lawton, aims to demonstrate the potential of permaculture to restore arid, desert environments. The project is located in Jordan's Jordan Valley, one of the most water-scarce regions in the world.

Key Features:

1. Water Harvesting Techniques:
   - The project employs swales, infiltration basins, and keyline design to capture and

store scarce water resources. These techniques have transformed the desert landscape, enabling plant growth and soil restoration.

- The use of wicking beds and other water-efficient gardening methods helps maximize water use and ensure plant health.

2. Soil Building:

- Organic matter, composting, and mulching have been used to build soil fertility in previously barren areas. This has created a fertile environment for a diverse range of plants.

- Leguminous plants are used to fix nitrogen in the soil, improving soil structure and fertility.

3. Community Involvement:

- The project involves local communities in learning and applying permaculture techniques. This fosters local knowledge, participation, and long-term sustainability.

- Educational programs and workshops are conducted to teach sustainable farming practices and promote environmental stewardship.

Impact:
The Greening the Desert Project provides a practical model for land restoration in arid regions. It demonstrates the potential of permaculture to address desertification, improve food security, and enhance community resilience.

Example 5: Tamera Ecovillage, Portugal

Location: Alentejo Region, Portugal

Overview:

Tamera Ecovillage is an intentional community in Portugal that integrates permaculture principles with social and spiritual aspects of sustainable living. Founded in 1995, Tamera aims to create a self-sustaining and regenerative community model.

Key Features:

1. Water Retention Landscapes:
   - Tamera employs ponds, swales, and other water retention features to capture and store rainwater. These water systems ensure a reliable water supply, support agriculture, and create habitats for wildlife.
   - The water retention landscapes have transformed the dry, degraded land into a lush, productive environment.

2. Renewable Energy:
   - Solar energy is extensively used for power, heating, and cooking, reducing the community's reliance on fossil fuels. The

community also explores other renewable energy sources, such as wind and biogas.

- Energy-efficient buildings and passive solar design help minimize energy consumption.

3. Diverse Ecosystems and Agroforestry:

- The community's design includes food forests, vegetable gardens, and integrated animal systems that enhance biodiversity and resilience. Perennial plants and polycultures are used to create sustainable and productive systems.

- The community practices natural pest control, soil building, and composting to maintain healthy ecosystems.

4. Social Permaculture:

- Tamera emphasizes community building, conflict resolution, and social sustainability. The community has developed various social structures and practices to foster harmony, cooperation, and resilience.

- Education and outreach programs are conducted to share knowledge and inspire others to create sustainable communities.

Impact:
Tamera Ecovillage demonstrates the holistic application of permaculture principles, integrating ecological sustainability with social and spiritual well-being. It serves as a model for intentional communities worldwide, showcasing the potential for creating regenerative and resilient societies.

Conclusion

These real-life examples of permaculture gardens illustrate the transformative potential of permaculture principles in various contexts. From urban homesteads to desert greening projects, permaculture can be adapted to diverse environments, creating resilient, productive, and sustainable systems. These examples provide valuable lessons and inspiration for anyone interested in adopting permaculture practices, proving that sustainable and regenerative agriculture is achievable in any context. By learning from these successful projects, gardeners and communities can embark on their own permaculture journeys, contributing to a more sustainable and resilient world.

Conclusion

Permaculture gardening represents a holistic and sustainable approach to agriculture and land management, integrating principles of ecology, design, and community living. Throughout this comprehensive guide, we've explored various aspects of permaculture, from its foundational principles and historical development to practical applications in garden design, soil health, water management, plant selection, and community building. The following key points encapsulate the essence and potential of permaculture gardening.

The Essence of Permaculture

Definition and Principles:
Permaculture, derived from "permanent agriculture," is a system of design that emphasizes sustainable practices modeled on natural ecosystems. Its core principles—such as observing and interacting with the environment, capturing and storing energy, and producing no waste—guide practitioners in creating regenerative and resilient systems.

History and Evolution:
Originating in the 1970s with the pioneering work of Bill Mollison and David Holmgren, permaculture has evolved into a global movement. It has expanded beyond agriculture to include sustainable living, community design, and ecological

restoration. The ongoing development of permaculture practices reflects the adaptability and relevance of this approach in addressing contemporary environmental challenges.

Practical Applications

Getting Started:
Understanding your space, observing and assessing your garden area, and considering climate and microclimates are critical initial steps. These activities lay the groundwork for effective and context-specific permaculture designs.

Design Principles:
Designing for sustainability involves applying permaculture principles like zoning and sector analysis, emulating natural patterns, and integrating multifunctional elements. This strategic planning ensures that the garden ecosystem is efficient, productive, and resilient.

Soil Health:
Healthy soil is the foundation of a successful permaculture garden. Techniques such as composting, mulching, and using cover crops enhance soil fertility and structure. Soil testing and amendments tailored to specific soil types further support plant health and productivity.

Water Management:
Effective water management is crucial in permaculture. Practices such as rainwater harvesting, greywater systems, and innovative irrigation techniques like drip irrigation and swales optimize water use, conserve resources, and support plant growth.

Plant Selection:
Choosing the right plants involves considering native and adaptable species, companion planting, and the balance

between perennial and annual plants. Integrating diverse plant guilds and polycultures enhances biodiversity, resilience, and productivity.

Garden Layout and Design:
Thoughtful garden layout and design incorporate elements like natural patterns, plant guilds, and multifunctional spaces. Creating edible landscapes and integrating food forests combine beauty and functionality, enriching both the environment and the gardener's experience.

Pest Management and Biodiversity:
Encouraging beneficial insects, implementing organic pest management strategies, and practicing crop rotation and diversity prevent soil depletion and enhance ecosystem health. These practices reduce reliance on chemical inputs and promote natural balance.

Tools and Resources:

Essential tools, both basic and specialized, facilitate efficient gardening. Investing in quality tools and learning proper maintenance techniques ensure longevity and effectiveness. Additionally, continuous learning through books, websites, courses, and community engagement enriches the gardener's knowledge and skills.

Real-Life Examples and Community Impact

Case Studies:
Real-life examples, such as Zaytuna Farm in Australia, the Urban Homestead in the USA, Krameterhof in Austria, the Greening the Desert Project in Jordan, and Tamera Ecovillage in Portugal, showcase the transformative power of permaculture. These case studies highlight the adaptability of permaculture principles to diverse environments and the tangible benefits of sustainable practices.

Interviews with Experienced Gardeners:
Insights from seasoned permaculture practitioners offer practical advice, inspiration, and wisdom. Their experiences underscore the importance of observation, experimentation, and community involvement in successful permaculture gardening.

Personal and Environmental Benefits

Personal Benefits:
Permaculture gardening offers numerous personal benefits, including improved physical health, mental well-being, and a deeper connection with nature. It provides a sense of accomplishment, self-sufficiency, and the joy of nurturing living systems.

Environmental Impact:
Permaculture contributes to environmental sustainability by enhancing biodiversity, conserving resources, and restoring

ecosystems. It mitigates the impacts of climate change, promotes soil health, and supports sustainable food production.

Moving Forward

Continuous Learning:

The journey of permaculture gardening is one of continuous learning and adaptation. Engaging with further reading, attending workshops, and connecting with the permaculture community are essential for staying informed and inspired.

Community and Global Impact:

Permaculture's principles and practices have the potential to address global environmental challenges. By adopting permaculture, individuals and communities contribute to a larger movement towards sustainability, resilience, and ecological harmony.

In conclusion, permaculture gardening is more than just a method of growing food—it's a comprehensive approach to living sustainably and harmoniously with the earth. By embracing permaculture principles, gardeners can create productive, resilient, and beautiful spaces that support both human and ecological well-being. The knowledge and practices shared in this guide provide a foundation for anyone interested in embarking on their permaculture journey, fostering a deeper connection with nature and contributing to a sustainable future for all.

www.ingramcontent.com/pod-product-compliance
Lightning Source LLC
Chambersburg PA
CBHW071909210526
45479CB00002B/344